Creative Placemaking

This book makes a significant contribution to the history of placemaking, presenting grassroots to top-down practices and socially engaged, situated artistic practices and arts-led spatial inquiry that go beyond instrumentalising the arts for development. The book brings together a range of scholars to critique and deconstruct the notion of creative placemaking, presenting diverse case studies from researcher, practitioner, funder and policymaker perspectives from across the globe. It opens with the creators of the 2010 White Paper that named and defined creative placemaking, Ann Markusen and Anne Gadwa Nicodemus, who offer a cortically reflexive narrative on the founding of the sector and its development. This book looks at vernacular creativity in place, a topic continued through the book with its focus on the practitioner and community-placed projects. It closes with a consideration of aesthetics, metrics and, from the editors, a consideration of the next ten years for the sector.

If creative placemaking is to contribute to places-in-the-making and encourage citizen-led agency, new conceptual frameworks and practical methodologies are required. This book joins theorists and practitioners in dialogue, advocating for transdisciplinary, resilient processes.

Cara Courage is an arts and placemaking academic and practitioner and is Head of Tate Exchange, Tate's programme and spaces dedicated to socially engaged art and the role of art in society. Her book, *Arts in Place: The Arts, the Urban and Social Practice* (2017), presents case-study research on social practice placemaking. Cara has also completed a project as Research Adjunct on the metrics of creative placemaking with Thriving Cities, an initiative of University of Virginia's Institute for Advanced Studies in Culture, and continues her own social practice arts in place projects.

Anita McKeown is an interdisciplinary artist, curator and researcher working in creative placemaking, Open Source Culture/Technology and STEAM (science, technology, engineering, arts and mathematics) education. Anita works for Art Services Unincorporated (ASU), an itinerant strategic platform which co-creates local-scale interventions that are context-responsive and ecologically sensitive, arising from extended relationships with people and place. ASU's interventions are underpinned by the permacultural resilience framework and practical toolkit, a critical praxis for *Creative Placemaking*, trialled in London, Dublin and New Mexico (2008–15). Anita's current research includes *codesres: Co-designing for resilience in rural development through peer-to-peer networks* and STEAM.

Routledge Studies in Human Geography

This series provides a forum for innovative, vibrant, and critical debate within Human Geography. Titles will reflect the wealth of research which is taking place in this diverse and ever-expanding field. Contributions will be drawn from the main sub-disciplines and from innovative areas of work which have no particular sub-disciplinary allegiances.

Place, Diversity and Solidarity
Edited by Stijn Oosterlynck, Nick Schuermans and Maarten Loopmans

Towards A Political Economy of Resource-dependent Regions
Greg Halseth and Laura Ryser

Crisis Spaces
Structures, Struggles and Solidarity in Southern Europe
Costis Hadjimichalis

Branding the Nation, the Place, the Product
Edited by Ulrich Ermann and Klaus-Jürgen Hermanik

Geographical Gerontology
Edited by Mark Skinner, Gavin Andrews, and Malcolm Cutchin

New Geographies of the Globalized World
Edited by Marcin Wojciech Solarz

Creative Placemaking
Research, Theory and Practice
Edited by Cara Courage and Anita McKeown

For more information about this series, please visit: www.routledge.com/Routledge-Studies-in-Human-Geography/book-series/SE0514

Creative Placemaking

Research, Theory and Practice

**Edited by Cara Courage
and Anita McKeown**

Routledge
Taylor & Francis Group

LONDON AND NEW YORK

First published 2019
by Routledge
2 Park Square, Milton Park, Abingdon, Oxon OX14 4RN

and by Routledge
52 Vanderbilt Avenue, New York, NY 10017, USA

First issued in paperback 2020

Routledge is an imprint of the Taylor & Francis Group, an informa business

British Library Cataloguing-in-Publication Data
A catalogue record for this book is available from the British Library

Library of Congress Cataloging-in-Publication Data
A catalog record has been requested for this book

ISBN 13: 978-0-367-58693-5 (pbk)
ISBN 13: 978-1-138-09802-2 (hbk)

Typeset in Times New Roman
by Integra Software Services Pvt. Ltd.

Contents

Figures

Contributors

Luísa Alpalhão is a London- and Lisbon-based architect and artist and founding member of the architecture, art and design platform atelier urban nomads. Luísa has an MA in Architecture and Interiors from the Royal College of Art and is currently a PhD candidate at The Bartlett School of Architecture, University College of London, with a scholarship from Fundação para a Ciência e Tecnologia. Her research consists of developing a methodology for the making of participatory processes that can potentially become a pedagogical tool for the inhabitation and understanding of urban shared spaces.

Sarah Barns is a creative producer and researcher working across urbanism, placemaking and experiential media. She co-directs the Sydney-based media arts and design practice Esem Projects, and is a Research Fellow at the Institute for Culture and Society, Western Sydney University. Sarah has led the creation of over twenty permanent and temporary installations across Australia and New Zealand, and continues to advise public and private agencies in fields spanning public art, urban strategy and community engagement. Her research has been published across a range of journals including *Space & Culture, City Journal, Senses and Society* and *Architectural Review*, among others.

Tim Edensor teaches Cultural Geography at Manchester Metropolitan University and is currently a visiting scholar at Melbourne University. He is the author of *Tourists at the Taj* (1998), *National Identity, Popular Culture and Everyday Life* (2002), *Industrial Ruins: Space, Aesthetics and Materiality* (2005) and *From Light to Dark: Daylight, Illumination and Gloom* (2017) as well as the editor of *Geographies of Rhythm* (2010). Tim has written extensively on national identity, tourism, ruins and urban materiality, mobilities and landscapes of illumination and darkness. He is currently writing a book about stone in Melbourne.

Anne Gadwa Nicodemus is a leading voice in the intersection of arts and community development. Her company, Metris Arts Consulting, provides planning, research and evaluation services to reveal arts' impacts and help communities equitably improve cultural vitality. Recent Metris projects span a case study of how a creative space in Zimbabwe fosters activism to a planning process that integrates arts and culture into the work of a community development organization with 250 affiliates. Since 2012, Nicodemus has been recognized as one of the USA's 50 most influential people in non-profit arts in WESTAF's annual peer-nominated list.

Margo Handwerker is the Director of the Texas State Galleries and a researcher member of the M12 Collective. She is the co-author of *A Decade of Country Hits: Art on the Rural Frontier* (2014).

Sunil Iyengar directs the Office of Research & Analysis within the National Endowment for the Arts (NEA), an agency of the US government. He is responsible for overseeing studies, program evaluations and analyses about the value and impact of the arts. Among his accomplishments at the NEA has been establishing a research grants program– with a special track for experimental and quasi-experimental study designs – and a 'Research Labs' program fostering transdisciplinary partnerships between researchers and arts practitioners. He chairs a federal Interagency Task Force on the Arts and Human Development, has edited dozens of NEA research publications and data visualizations and has led the development of the National Archive of Data on Arts & Culture, a free data repository available at www.icpsr.umich.edu/icpsrweb/NADAC/.

Torange Khonsari is an academic and practitioner specialised in citizen-led city development. She co-founded the art and architecture practice Public Works (2004), an inter-disciplinary practice working in the threshold of participatory art, architecture, anthropology and politics that tests and implements the academic research undertaken at The Cass, London Metropolitan University. Khonsari is a Consultant on the Specialist Assistant Team (SAT) of the Mayor of London on community development and cultural curation in regeneration. She is Senior Lecturer at London Metropolitan University where she runs MA Design for Cultural Commons. She has been a Consultant to UN Habitat on sustainable development.

Anna Marazuela Kim is a Research Fellow of the Thriving Cities Lab at the Institute for Advanced Studies in Culture at the University of Virginia, where she advances research on the role of art and aesthetics in civic thriving. Kim is an art historian and cultural theorist who writes and lectures on our complex relation to images and their continuing ethical challenge, from Plato to the digital age. She is the recipient of numerous fellowships and awards and member of several international research groups. Currently she is a Visiting Research Fellow at the Institute for Advanced Studies at University College London.

Michael Landzelius is Director of the Urban Safety and Societal Security Research Center (URBSEC) at University of Gothenburg and Chalmers University of Technology. He is Associate Professor, positioned at the Department of Sociology and Work Science, University of Gothenburg. He did his PhD in Conservation of Built Environments, with an approach shaped under a post-graduate year studying Geography at University of California Berkeley and Syracuse University. Michael also spent two years as a postdoc in Geography at the University of Cambridge. In addition to his present role as URBSEC Director, he is Project Leader in a four-year research project entitled Reconciliatory Heritage: Reconstructing Heritage in a Time of Violent Fragmentations.

Ann Markusen is Principal of Markusen Economic Research (annmarkusen.com) and Professor Emerita at the University of Minnesota. Her publications include *Creative Capital Artists Look Back* (2016); *City Creative Industry Strategies* (2012); *Creative Placemaking* (2010); *Native Artists; Crossover: How Artists Build Careers across Commercial, Non-Profit and Community Work* (2006); and *Artists' Centers* (2006). Markusen, who has a PhD in Economics from Michigan State University, was Professor at Minnesota, Rutgers, Northwestern, California Berkeley and Colorado Universities; Fulbright Lecturer, Brazil; White Professor-at-Large, Cornell University; Bousfield Distinguished Professor, University of Toronto; Harvey Perloff Visiting Chair, UCLA; UK Fulbright Distinguished Chair, Glasgow School of Art; and is currently a member of the National Advisory Board Strategic National Arts Alumni Project.

Shauta Marsh is a co-founder of Big Car Collaborative, formed in 2004. From 2011 to 2015 she was Executive Director of the Indianapolis Museum of Contemporary Art (iMOCA). There she curated and/or organized more than forty exhibitions with artists including LaToya Ruby Frazier, Tony Buba, Trenton Doyle Hancock and Richard Mosse. She returned to Big Car in March 2015 as a Commissioning Curator and Program Director for its new headquarters, *Tube Factory* artspace. Since opening the 12,000 square foot museum/community center hybrid, she has worked with artists Carlos Rolón, Jesse Sugarmann, LaShawnda Crowe Storm, Mari Evans, Pablo Helguera, Scott Hocking and many others.

Steve Millington is a Director of the Institute of Place Management and a Senior Lecturer in Human Geography at Manchester Metropolitan University. Working in partnership with key industry stakeholders and local centres, he is a co-investigator on two major research projects analyzing town centre change and development in the UK: ESRC-funded High Street UK2020 and Innovate-funded Bringing Big Data to Small Users. Steve is co-editor of two edited collections, *Spaces of Vernacular Creativity* (2009) and *Cosmopolitan Urbanism* (2005). He has published research on several facets of placemaking, including lighting and place, and the relationships between football clubs and their localities. Formed in 2006, the Institute of Place Management is the international professional body that supports people committed to developing, managing and making places better. The concept for the Institute was developed by the Manchester Metropolitan University and the Association of Town Centre Management in the UK.

Aditi Nargundkar Pathak is an architect and urban designer. She leads the placemaking initiative The Urban Vision, enabling the use of public art in plazas and designing human-centric streets and innovative plazas in Indian cities. She has led and completed multiple demonstration projects creating small social spaces which have changed the safety, aesthetics and use of the city's public spaces. She is a visiting faculty in various colleges of architecture and urban planning in Mumbai.

Stephen Pritchard is a Fine Art Tutor at Northumbria University. He recently completed an AHRC-funded research-based PhD entitled *Artwashing: The Art of Regeneration, Social Capital and Anti-Gentrification Activism*. He is an art historian, critical theorist, activist, writer, curator and community artist. His interdisciplinary approach to research is grounded in post-critical ethnography, radical art history, Frankfurt School Critical Theory and Critical Urban Theory. He has presented papers internationally, lectures widely and has had several journal articles published to date. He also was commissioned by *The Guardian* to write an article entitled 'Hipsters and Artists are the gentrifying foot soldiers of capitalism' in 2016. Stephen is also a member of the Movement for Cultural Democracy and Artists' Union England, and is a founding member of Isla99, Artists Against Social Cleansing and Socially Engaged and Participatory Arts Network. He has worked in the arts since 2007 and founded the community arts organisation dot to dot active arts CIC in 2013.

Peter Rundqvist has for many years been part of the City Council administration of Gothenburg, developing and leading different EU-financed sustainable urban development projects. He is a sociologist specialised in the field of art and culture related to social cohesion and migration in contemporary urban development processes.

Dominic Walker is a cultural geographer at Royal Holloway, University of London. His research explores the interface between art, politics and science in contemporary environmental humanities discourse. Walker presented papers at the Royal Geographical Society (2015, 2016, 2018) and American Association of Geographers (2016) annual international Geography conferences, alongside guest lectures in Pittsburgh (2015) and Exeter (2014, 2015). He has collaborated with several international artists, and was a Visiting Research Fellow in Carnegie Mellon University's School of Art (2015) and the Center for PostNatural History (2015). He has published in *Society and Space* (2015) and has a further two papers awaiting submission to the journals *GeoHumanities* and *Area*.

Jim Walker is CEO of Big Car Collaborative, an Indianapolis, Indiana-based nonprofit social practice art and placemaking non-profit. Walker serves as lead artist on *Spark Placemaking*, a Big Car program bringing engagement-based public programming and people-focused design to communities. He also leads Big Car's work to utilize cultural strategies to support equitable revitalization in long overlooked neighborhoods south of downtown

Indianapolis. Walker – who received his MFA from Warren Wilson College in Asheville, N.C. – has worked as a teacher, journalist, designer, and public artist. He also currently teaches in the University of Indianapolis Social Practice Art and Placemaking graduate program.

Joshua J. Yates is a cultural sociologist and social entrepreneur whose work bridges the worlds of academic theory and social practice. He is currently Research Director of the Thriving Cities Lab, an initiative of University of Virginia's Institute for Advanced Studies in Culture, where his scholarly work focuses on the changing paradigms of civic life in 21st-century urban contexts, and Chief Executive Officer of the Thriving Cities Group, a non-profit organization equipping cities to cultivate their civic capacity and civic infrastructure through community-centred data, technology and advice.

Preface

This book has arisen from a transatlantic conversation between eponymous creative placemaking conference sessions at the Royal Geographical Society, Exeter, 2015, and the Association of American Geographers, San Francisco, 2016. These sessions worked to unpick the notion of creative placemaking and offered a practitioner-led critique and examples of practice in conversation with an academic- and research-based discourse. As placemaking practitioners and researchers, we both felt that the arts-driven placemaking sector had reached a moment in both maturity and breadth where it demanded critique and a deeper understanding of practice, requiring in turn a meaningful and dynamic dialogue between theorists and practitioners. If creative placemaking is to contribute to places-in-the-making (Silberberg, 2013) and encourage citizen-led agency, new conceptual frameworks and practical methodologies will be required, advocating transdisciplinary, resilient processes and new models of theory and practice. This book aims to be part of that sector development.

Creative placemaking is often addressed as a subset of placemaking and is commonly, and this book would argue, reductively, recognised for the instrumentalised potential of art to contribute to regeneration and revitalisation. Such an approach has little engagement with the heritage of non-object arts, design or architectural practices, those situated interventions and durational practices that have emerged over fifty years of 'arts in place' and in much placemaking spanning both the social and public realms. This book works then to interrogate a populist or sector vernacular definition and understanding of creative placemaking and to extend its definitions and understandings through both academic and practitioner voices. As a still-nascent field, emerging from a US policy platform, creative placemaking is still evolving, yet enacted at a global scale. A review of key funders from 2010 to 2015 by co-editor Anita McKeown revealed that the majority of creative placemaking projects were artists' live-work spaces, cultural quarters, landmark arts centres or monumental public artworks. This focus does little to advance the field or present emergent praxis, which is systemic in its approach. The book aims to address this deficit by representing a range of practices across themes that are pertinent to the social or public realms and to signal progressive changes to future challenges. A range of scholars and artist-scholars present socially practiced, co-produced and citizen-led placemakings as an inside-out response rather than simply a bottom-up need or desire or top-down imposition, with artists, participants and a range of creatives and other professions forming ecologies of practice.

The chapters in this book don't all agree with each other – it is not our purpose to form a consensus, but to give a platform to a diverse cohort of voices across the maturing creative placemaking sector, and to prompt the reader into their own critique, reflection on their own practice and position in the sector. We embrace differing perspectives and opinions in the construction of our conference session and this book, each chapter in dialogue with others and inviting further research and debate, which we hope will have a life beyond the publication. We are honoured too to have the reflexive voice from the founding of creative placemaking and from its practice, offering a mix of first- and third-person narrative, conversation and reflection, a reflection itself on the nature of a critiquing enquiry and of the prominence of the active placemaker, a 'thinking through doing' and of practice-based research. With professional practices in social practice curation (Cara), as an artist (Anita) and as placemakers (both), we took a curatorial approach to the conferences sessions, and have done so again for this book. This approach is discussed in the opening chapter.

Both UK and American English feature in this publication, recognising the voice of contributors and direct quotes from other texts, artists and project participants.

Reference

Silberberg, S. (2013) *Places in the Making: How placemaking builds places and communities.* MIT Department of Urban Studies and Planning [Online] Available at: http://dusp.mit.edu/cdd/project/placemaking (Accessed: 13 August 2015)

Acknowledgements

The editors would like to thank all the contributors for their dedicated work and inspired contributions to this book, as well as the team at Routledge. Thanks are also due to the Royal Geographic Society, UK, and the Association of American Geographers, USA, for giving an ongoing platform to bring arts into geographical contexts and organising the conferences that were the beginning setting of this publication. We would like to take this opportunity to acknowledge two of our valued session contributors, Marie Mahon, Ireland, and Eje Kim, Republic of Korea, who were unable to contribute to this publication. Furthermore, we would like to thank all those communities and artists referenced in this book for their invaluable contribution to questioning and extending creative placemaking practice.

Abbreviations

a2ru	Alliance for the Arts in Research Universities
AAF	American Architectural Foundation
AAG	American Association of Geographers
ASU	Art Services Unincorporated
AUD	Australian dollars
AUN	Atelier Urban Nomads
BHAAAD	Boyle Heights Alliance Against Artwashing and Displacement
BSC	Balfron Social Club
CAE	Critical Art Ensemble
CCI[s]	Cultural and Creative Industries
CEO[s]	Chief Executive Officer[s]
CLT[s]	Community Land Trust[s]
DOT	Department of Transportation
ERDF	European Regional Development Fund
GDNE	Gothenburg Development North East
HUD	Housing and Urban Development Department
IHRU	Instituto de Habitação e Reabilitação Urbana
iMOCA	Indianapolis Museum of Contemporary Art
LGBTQI+	Lesbian, Gay, Bisexual, Transgender, Queer, Questioning and Intersex
LQC	Lighter-Quicker-Cheaper
MACLA	Movimiento de Arte y Cultura Latino Americana
MICD	Mayors' Institute on City Design
MSB	Maritime Housing Board
NEA	National Endowment for the Arts
NGO[s]	Non-Governmental Organisations
NSW	New South Wales
ORU	Operações de Reabilitação Urbana/Urban Rehabilitation Operations
PwC	PricewaterhouseCoopers
RFP	Request for Proposals
RGS	Royal Geographical Society
RSPB	Royal Society for the Protection of Birds

RTR	Roman Road Trust
SEK	Swedish krona
SIAP	Social Impact of the Arts
SNAG	Southwark Notes Archive Group
STEAM	Science, Technology, Engineering, Arts and Mathematics
TRF	The Reinvestment Fund
URBSEC	Urban Safety and Societal Security Research Center
USCM	United States Conference of Mayors
V&A	Victoria and Albert Museum
VALI	Validating Arts and Livability Indicators

Introduction

Curating research, theory and practice

Cara Courage and Anita McKeown

The curatorial approach taken to the conference sessions this book has arisen from, and to the book itself, was a deliberative, and, as arts practitioners, intuitive act, orientated as we are towards an expansive understanding of the role of the curator and the curatorial praxis.

The arts sector has seen a curatorial turn that has widened its scope from art museum or gallery exhibition to include 'enabling, making public, educating, analysing, criticising, theorising, editing and staging' (Jackson, 2015, p.62) – the knowledge base of the curator has expanded and been de-centred, and multiple disciplines have been incorporated into its purview. We take this a stage further to include academic presentation across research dissemination platforms and the practitioner voice in the academy. Within our practice and academic fields (separated here only notionally), we see an 'entanglement of actors' (Jackson, 2015, p.15) rather than siloed or fixed-boundary positions and seek to engage academia in this praxis as an active concern. For us, the curatorial is a 'a more viral presence consisting of signification processes and relationships between processes and relationships between objects, people, places, ideas, and so forth, a presence that strives to create friction and push new ideas' (Lind, in Jackson, 2015, p.65).

Thus, our curation is co-produced, is collaborative and dialogical, as is our placemaking practice. With the abundance of content and channels of content communication within academia and the arts alone, the curatorial is a vital and critical approach to take for sense-making, dissemination and whether academic, researcher or practitioner, the expansion of our intellectual and creative endeavours. The curatorial imbricates the folding and unfolding of academic and practitioner knowledge and provides a critical framework to find and assess information. It requires traditional and information literacy, visual and critical literacy (Dale, 2014, p.203) and the critical faculty to add value to, rather than just share, information. This is curating of the cognitive value and the material in this book has been chosen to support the best writing and research undertaken in the creative placemaking sector and to make this work more available to those within and without the academy, especially, as a practitioner-led field, to the placemakers themselves.

Our co-organised conference sessions at the Royal Geographical Society (RGS) and American Association of Geographers (AAG) (see Preface) and this book manifest a curatorial framework of critically engaged practice and a 'mode of becoming' (O'Neill and Wilson, 2015, p.12) of theory, research and practice. Responding to the conferences themes ('Geographies of the Anthropocene' at RGS, and at AAG, 'Thriving in a Time of Disruptive in Higher Education', 'International Geography and Urban Health Symposium' and 'Symposium on Physical Geography and Challenges of the Anthropocene'), we devised our thematic response of galvanising, continuing and re-invigorating the arts-led conversation, our mode of enquiry for the sessions and learning objectives and plenary questioning. The sessions were formed from an open call for papers as well as targeted approaches to our trusted networks. Submitted abstracts were put through a process of filtration, evaluation and assemblage, checked against the theme and our session-learning objectives. The disparate sources of information that is a portfolio of papers were ordered to present a coherent narrative arc across the theme; for example, moving from a global to a local perspective or moving from the theoretical to the practice-based enquiry.

This process, informed by our curation as artistic practice, utilises 'form' as a means to advance understanding and speakers were offered guidance in advance on content in this regard. When in the conference room, the sessions were a space for active engagement with experts and peers between speakers and delegates, resulting in a mutual gaining of insight and embedding of new thinking. The composition of the narrative arc curated the panellists' presentations as a series of landing points from which to navigate the field. This served to create an experience in which the space between the contributions informed understanding as much as what was presented.

Through traversing geographic, intellectual and infrastructural constructs, the sessions' curation sought to encourage a diverse landscape to create an interdisciplinary ecology that facilitated the cross-pollination of knowledge. This in turn encouraged and supported our understanding of creative placemaking as a collaborative, participatory process and taking an ecological approach to theoretical and practical framing and questioning. Seating plans of plenary sessions were chosen for best fit to the learning objectives, and ranged from conventional classroom setting to cabaret, world café, conversation circle. Interdisciplinarity has been a constant mode of enquiry – 'not only a matter of going beside the disciplines but of breaking them' (Jackson, 2015, p.6) – as has been the curatorial conversation of the closing plenary or discussant. To this end, our curatorial approach through this publication further endeavours to encourage the reader to select sections and chapters at will, co-curating their own exploration of the contributions to develop further understanding and questioning.

The curation of the conference sessions sought to engage with the vital conversations that we as editors, through our own research on placemaking and earlier critiques of cultural regeneration, had identified as problematic for the creative placemaking sector, which, principally at the time of writing, revolved around issues of gentrification, participation and exclusion. Our purpose behind curating research, theory and practice is to offer an approach that can structure and demonstrate a multiplicity of positions, models and voices at their intersection (O'Neill and Wilson, 2015, p.12). Topics and themes create a constellation that can generate reflection, debate and the extension of both research and practice processes. This operates on two levels: the choosing and framing of the topic, and the choosing of its components and their presentation, 'which (together via the programming) enable and enhance reflexive dialogues among audiences and participants' (Nelund, 2015, p.174). Throughout this process, our own reflective practice mined the RGS sessions to extract common themes, questions and avenues for further inquiry that informed our second set of sessions at AAG the following year.

This curation is further manifest within the book's *Ecologies* organisational structure, which references a conversation deeply embedded in creative place-making's heritage: the need for an ecological approach as identified in the work of Stern and Seifert (2006) within the Social Impact of the Arts (SIAP) at Penn State. In a radical pairing, Joan Shigegawa, Head of the Rockerfella Foundation (soon to be Deputy Chair to Rocco Landesman's Chair of the National Endowment for the Arts (NEA)), negotiated an innovative collaboration between SIAP and The Reinvestment Fund. This culminated in an influential report – *Culture and Urban Revitalisation* (2006–8), an important in-depth critique of *The Rise of the Creative Class* (Florida, 2002), which recommended the need for an 'ecological' approach to including the arts within 'urban revitalisation'. Shigekawa's relationship to this partnership and report, an important influence on the early development of creative placemaking, is evident in the curatorial organisation of the book.

Further, the use of ecologies as metaphor and formal construct presents not only the need for an ecological systemic approach to actualise creative place-making's generative potential but references the need for dynamic practices that transcend the bi-polarities of top-down or bottom-up, a key discontent within the field, and which are necessary to devolve power, encourage self-organisation and agency and integrate citizens' existing placemaking practices. Resilient and co-produced instances of creative placemaking are moving towards a processual open source approach (Silberberg, 2013) and have become more common since our exploration of the field began in 2015. Early drivers of the sector, the NEA's and ArtPlace's reflection on their own processes now sees support shifting from capital projects, such as the monumental and plaza or waterfront development, towards a broader transference of power from external local or centralised authority and towards opportunities to develop agile resilient practices.

Section 1 – Evolving Ecologies

The book's curation seeks to present advancement, and as such the first section, Evolving Ecologies, opens with two reflective chapters that reflect on the field through revisiting earlier scholarship. The four scholars' willingness and enthusiasm to rethink earlier research and present the long view initiates a grounded exploration of creative placemaking, our first aim for the book. The theoretical and practical concerns that the book presents move away from a traditional placemaking foci often manifest within urban development as the waterfront, plaza or the market place towards arts-led socially engaged processes that encourage a deeper level of public participation and agency. The opening section sets the scene for this objective by presenting key concerns that have been present within the field since its inception, yet acknowledges that despite progress, creative placemaking is a process that is still evolving. Ann Markusen and Anne Gadwa Nicodemus, the authors of the original White Paper (2010) commissioned by Rocco Landesman, Chair of the NEA, review close to a decade of work, designed by the Obama-appointed NEA administration. This foundation is strengthened by Edensor and Millington's reflections on their edited volume *Spaces of Vernacular Creativity: Rethinking the Cultural Economy* (2009). Emerging as Landesman was coining the phrase 'creative placemaking', Edensor and Millington presented a critical response to predominant discussions on creative geographies. As a companion piece to Markusen and Gadwa's text, Edensor and Millington's chapter serves to contribute to an understanding of the landscape that informed our initial concerns: the potential of creative placemaking and the limitations posed by a focus on economic impact and how this in turn shapes regeneration policy.

The second objective of the book was to extend definitions and the understanding of creative placemaking, and the chapters that follow Section 1 offer a rich breadth and depth of projects that traverse urban design and art heritage, theory and practice, planning and policy and cultural and place heritage and politics. Many of the scholars and practitioners contributing to this collection are engaged in a critical process of creative placemaking as well as moving beyond the confines of their discipline and its training to interrogate the issues that have led to many of the criticisms of the practice and sector.

Section 2 – Dialogical Ecologies

The three chapters presented in Section 2, Dialogical Ecologies, illustrate the book's aims by initiating a conversation between theory and practice from three perspectives: a geographer, an artist-scholar and an architect. Collectively, these chapters simultaneously highlight the potential of creative placemaking and its discontents from within three distinct geo-political contexts, reaffirming the need for an overarching ethos that allows for non-formulaic context-responsive and adaptive approaches. Within these chapters the authors present key criticisms of creative placemaking that pertain to issues of place attachment and belonging, present in discussions of creative placemaking from early in its inception. Creative practice in the broadest sense offers avenues for understanding and expressing place identity through articulation of psychosocial processes including comprehension of self and other, inclusion and exclusion in processes and systems that affect feelings of belonging and, as Bedoya (2013) has discussed, disbelonging.

Walker's reflection on a collaborative interdisciplinary citizen-led social practice project, *Wrecked! On the Intertidal Zone* (2014–6), shines a light on the
value of the arts to contribute beyond economic development. Through situated
practice, *Wrecked!* brought together multiple actors – local arts groups, citizens,
Arts Catalyst and Critical Art Ensemble – to produce social and cultural capital
as a response to water pollution and high shipping volume in the Thames
Estuary, UK. Barns raises awareness of the instrumental logic behind the support
of culture-led projects in Australia through government finance and urban
regeneration projects, and cites this as leverage for temporary programming.
This segues into an exploration of the renowned 'Lighter, Quicker, Cheaper'
(Reynolds in MacIver, 2010) process (advocated for by the placemaking agency
Project for Public Spaces) within the final chapter in this section through projects
in Mumbai by Patak, the Director of Urban Vision, along with a team of urban
innovators, architects, artists and urban designers. Urban Vision's consideration
of a Western concept highlights issues prevalent in all three sections: the need for
awareness and sensitivity to local contexts when transferring models and working in contested arenas.

Section 3 – Scalable Ecologies

Scalable Ecologies positions perspectives of scale in dialogue, bringing together
three contributors who address issues of location through rural, the peri-suburban
to the urban; issues of practice from the individual to municipal level; organisational issues within small and larger scale actions and approaches to creative
placemaking. Handwerker, a researcher and member off M12 arts collective,
assesses creative placemaking from a rural perspective informed by critical and
historical reflections augmented by a practice that is place-based and by the
realities and complexities of rural life in the United States. Handwerker raises
awareness of challenges to creative practices in a rural context and how creative
placemaking can dilute limited funds from the arts to social services and private
development, rather than being a means to increase resources and embed the arts
more concretely in all aspects of placemaking.

Landzelius and Rundqvist chart the evolution of a project in a peri-urban
context, the north-east of Gothenburg, which explored an assumption that
Cultural and Creative Industries (CCIs) could serve as an instrument for the
city's integration and economic development policies. The chapter presents the
shift from cultural policy driven by a socially orientated state-level system
towards an entrepreneurial neoliberal approach, while illustrating the difficulties and conflicting ideologies and social requirements at play within public
sector-initiated projects within 'vulnerable, economically weak and socially
segregated areas'.

The section concludes with a conversation undertaken on a research road trip
between Big Car Collaborative founders Walker and Marsh as they visited North
American rust-belt cities. Now in its fifteenth year as a socially engaged art and
placemaking organisation, Big Car is an integral part of a community in the

urban core of Indianapolis, Indiana. Literally travelling through ideas, experiences and a physical landscape that is adopting creative placemaking solutions as they drove, Big Car present pertinent aspects for creative placemaking practitioners.

Section 4 – Challenging Ecologies

Challenging Ecologies presents the perspective of three scholar-practitioners testing their fields' conventional practices. These perspectives offer important insights into a critical reflection of standard practices that informs both the field of creative placemaking and the education and training of those working professionally in disciplines that are engaged in the practice. Concerns around ownership and authorship are dealt with in different ways by each author yet underpinning their contributions is the recognition that creative placemaking is not the preserve of the 'professional placemaker'; instead they acknowledge the necessary and long-standing role multiple individuals contribute to the practice of placemaking that is creative and occurs every day and in many ways.

Khonsari's introduction to the role of temporary spatial objects/architectures considers how such territorial interventions can facilitate civic use and social empowerment as a tactical approach to resist privatised enclosures in urban areas. Using a comparative study, Khonsari presents the historical context of temporary spatial objects/architectures: the Soviet *Agitational Propaganda Vehicles* (*Agitprop* trains, 1920) and *The Fun Palace* by Joan Littlewood and Cedric Price (1960) in conjunction with two contemporary London case studies. Khonsari's counterpoint highlights the 'problematics of control and power' and the need for a spatial practitioner to have critical understanding of power relationships and dynamics within neoliberalism.

Following on from Khonsari's polemic, Pritchard suggests that creative placemaking perpetuates the gentrification associated with Creative City and Creative Class models, arguing that creative placemaking enacted via state and local authority policy and corporate partnerships, integrates art, community and economic development, and as such is a neo-liberal tool. As a rebuttal to this argument Pritchard advocates the process of 'place guarding' (collective acts of protecting existing people and places) as a way of artists resisting artwashing (the use of art as a veneer or mask for corporate or state agendas) (O'Sullivan, 2014), which, he argues, is embedded within creative placemaking. Alpalhão's pertinent reflection on participation and apathy within *Outros Espaços* (*Other Spaces*) (2014–15) highlights the complexity and pitfalls of participatory civic engagement within multi-agency regeneration.

Section 5 – Extending Ecologies

The closing section, Extending Ecologies, offers signposting to the type of thinking that is taking the creative placemaking sector into its next era. Kim and Yates' chapter builds on current understandings of the complexity of cities and their significance for the future. Increasingly, there is recognition of the need for an ecological model of cities that acknowledges an ecosystem with living and non-living organisms co-existing in dynamic interactions. This poses challenges to conventional approaches to planning and management as such environments are complex, with their dynamism requiring conditions that allow for constant change and evolution if they are to thrive.

The final chapter returns the reader to creative placemaking's inception, as Iyengar, the NEA's Director of the Office of Research & Analysis, documents its journey and move to a Theory of Change model. Iyengar's reflection is an optimistic perspective on creative placemaking's evolution at practitioner, funder and administrative levels. It is an appropriate milestone on which to end the book, as creative placemaking nears completion of its first decade. Through a longitudinal consideration and evaluation of intended and unintended outcomes from practitioners, the NEA has listened, acknowledged and integrated the knowledge from those enacting creative placemaking in the field.

In this spirit, we now invite you to follow our own journey through this book, following our curated path or by devising your own, and hope that it both illuminates creative placemaking practices and processes and informs your own research, theory and practice in the same.

References

Bedoya, R. (2013). *Creative Placemaking and the politics of belonging and disbelonging*. World Policy, Arts Policy Nexus series May 13, 2013. Available at: www.worldpolicy. org/blog/2013/05/13/creative-placemaking-and-politics-belonging-and-dis-belonging. [Accessed: 25 March 2018].

Dale, S. (2014). 'Content curation: The future of relevance', *Business Information Review*, 31 (4): 199–205.

Davis, J. L. (2017). 'Curation: A theoretical treatment', *Information, Communication & Society*, 20 (5): 770–783.

Edensor, T., Leslie, D., Millington, S., & Rantisi, N. (Eds.). (2009). *Spaces of Vernacular Creativity: Rethinking the Cultural Economy*. New York: Routledge.

Florida, R. L. (2002). *The Rise of the Creative Class: And How It's Transforming Work, Leisure, Community and Everyday Life*. New York: Basic Books.

Jackson, G. (2015). 'And the question is. . .' in *Curating Research*, O'Neill, P. and Wilson, M. (eds.). London: Open Editions.

MacIver, M., (2010). *Eric Reynolds, Master of Low-cost, High Return Public Space Interventions in London and NYC*. Available at www.pps.org/blog/ericreynolds-master-of-low-cost-highreturnpublic-space-interventions-in-londonand-nyc/. [Accessed: 9 August 2010].

Nelund, S. (2015). 'Home works' in *Curating Research*, O'Neill, P. and Wilson, M. (eds.). London: Open Editions.

O'Neill, P. and Wilson, M. (2015). 'An opening to curatorial enquiry: Introduction to curating and research' in *Curating Research*, O'Neill, P. and Wilson, M. (eds.). London: Open Editions.

O'Sullivan, F. (2014). *The Pernicious Realities of 'Artwashing'*. [Online] Available at: www.citylab.com/housing/2014/06/the-pernicious-realities-of-artwashing/373289/ [Accessed 26 September 2016].

Potter, J. and Gilje, Ø. (2015). 'Curation as a new literacy practice' *E-Learning and Digital Media*, 2 (2): 123–127.

Silberberg, S. (2013) *Places in the Making: How Placemaking Builds Places and Communities*. MIT Department of Urban Studies and Planning [Online] Available at: http://dusp.mit.edu/cdd/project/placemaking (Accessed: 13 August 2015)

Stern, M. and Seifert, S. (2006). *From Creative Economy to Creative Society. Creativity and Change*. Philadelphia, PA: Social Impact of the Arts Project and The Reinvestment Fund.

Section 1

Evolving Ecologies

1 Creative placemaking

Reflections on a 21st-century American arts policy initiative

Ann Markusen and Anne Gadwa Nicodemus

Abstract

The US National Endowment for the Arts' *Our Town* initiative and its philanthropic counterpart, ArtPlace, emerged in 2010/2011 in response to a 1990s crisis in national arts funding and retreat into economic impact advocacy. We recount the Obama administration's arts policy leadership and their 'creative placemaking' approach. We analyze two ongoing challenges – diversity and displacement – addressing how to broaden participation by people and communities of color and avoid displacing low-income residents and small businesses. We review the indicators approach initially embraced by National Endowment for the Arts and ArtPlace for evaluation, exploring its conceptual fuzziness and challenges of spatial scale, suitable data, and exogenous counterforces. We recap the spread of the creative placemaking ethos, celebrating its invitation to artists and arts organizations to use their artistic creativity for the good of their communities.

Introduction

The creative placemaking rubric frames a decade of new work by the US National Endowment for the Arts (NEA)[1] and its companion philanthropic consortium, ArtPlace.[2] Both organizations fund locally initiated community and economic development projects with arts and culture at their core. They also fund knowledge-building exercises (the NEA) and field scans (ArtPlace.) Designed by the President Obama-appointed team at the helm of the NEA, the initiative's intent has been to celebrate and embed arts and cultural capacity in neighborhoods and communities, contributing to prosperity (jobs, small-business income), social wellbeing, public safety, and stability. Both funders encourage arts organizations and artists to preserve and enliven places by using their visual, musical, speech, writing, and acting skills for and in conjunction with larger publics. Since the publication of the NEA-commissioned White Paper *Creative Placemaking* (Markusen and Gadwa, 2010), interest in this approach has spread to many countries and their small as well as large cities. For example, the Czech Republic's Prague/Pilsen Year of Culture adopted creative placemaking as its overarching theme (Markusen, 2014; Markusen and Gadwa Nicodemus, 2015), and the South Korean national arts agency translated the original *Creative Placemaking* document into Korean. In this reflective essay, we employ a political-economy point of view, charting the challenges that almost eliminated the NEA in the 1990s, the subsequent advocacy shift towards the economic impact of the arts, and the emergence of the NEA's *Our Town*[3] initiative in 2011. We take policy initiatives, their rationales, and their implementation seriously, placing them center stage.

Creative placemaking deliberately conflated the creative class approach of Florida (2002) with decades of progressive community-culture-based placemaking, offering US advocates of greater public engagement in arts creation and presentation new channels for accomplishing this. The initiatives, and the financial and organizational power behind them, prompted productive debates about the roles of artists and arts organizations as well as the potential for coalitions and their interventions to gentrify neighborhoods and in doing so displace long-time lower-income residents and the commercial activities that served them. We probe the issues of displacement and racism in detail as unanticipated challenges for communities and funders, citing pertinent examples and offering policy solutions.

In a following section, we address the considerable evaluative challenges for funders and placemakers, especially given cultural diversity and placekeeping priorities. Part of the 'creative' in creative placemaking is precisely this – that the fruits may emerge very differently than initial intentions, just as we see in funded science projects all the time. Engaging partners and stakeholders in conversations about what success looks like and for whom can be useful in helping people move beyond fuzzy concepts into what they actually hope to accomplish. Creative placemaking evaluation sometimes confronts difficulties accounting for conflicting agendas of participants: the development industry (often quite powerful locally), artists, art organizations, neighbourhood leaders, and elected officials regarding creative placemaking projects. However, evaluators face steep challenges surrounding good, grounded research that can produce comparative data across localities. We also discourage evaluation efforts that attempt to winnow out winners from losers.

Overall, the NEA *Creative Placemaking* and companion ArtPlace initiatives have significantly improved the general public's view of the role of artists and arts capabilities to serve their communities while making substantial contributions to community stabilization and cultural engagement in many places. The two funders are now placing more emphasis on long-term integration of arts and culture into comprehensive community development. They are also encouraging greater grassroots participation as a priority in their criteria for awarding funding. We remain hopeful that debates about diversity, displacement, and evaluation are generating positive changes in the design of the funding streams and outcome assessments. Certainly, following a decade of these initiatives, creative placemaking efforts are incorporating many more people of all types in the debate over who, how, and why we should support arts and cultural activity. Will creative placemaking disappear if the NEA is defunded and ArtPlace gracefully shuts its doors at the end of its ten-year tenure? We argue that it will not, but that its varieties and protagonists have and will continue to morph. We are optimistic about its long-term contributions and hopeful for greater gains for long discriminated-against racial and immigrant groups, for greater public participation, for a flourishing of participatory arts, and as Bedoya (2013, 2014) demands, 'placekeeping' as enduring features.

Creative placemaking as cultural policy shift

Beginning in the early 1990s, US cultural policy endured a Congressional rebellion that nearly eliminated its National Endowment for the Arts, a modestly funded federal agency that supported individual artists and arts organizations directly and passed a significant share of its funding to state- and local-government arts agencies. Several NEA-funded performance and visual artists were responding to the AIDS crisis in their work (Killacky, 2011, 2014), prompting a Republican Congress to savage the NEA budget. Corporate and private philanthropic support for the arts declined as well (Kreidler, 1996). Bill Ivey, President Clinton's Chair of the NEA, argues that this implosion was also due to elitist attitudes and aesthetics that had driven NEA funding away from popular culture (Ivey, 2008).

The response of the arts community was timid. Rather than defend the artists in question, the national arts advocacy organization, Americans for the Arts (AFTA),[4] and its state-level counterparts turned to economic arguments. AFTA continues to publish studies showing how arts spending generates income and jobs for states, localities, and the nation (e.g. Americans for the Arts, 2016). Florida's *The Rise of the Creative Class* (2002) generated considerable interest among journalists and state and local public officials, despite forceful critiques of the book's theoretical and empirical quality (Markusen, 2006; Markusen et al., 2008). *Gifts of the Muse: Reframing the Debate about the Benefits of the Arts* (McCarthy et al., 2004) championed the arts' intrinsic contributions. Complementary pioneering work offers evidence and methods for estimating the arts' intrinsic value to participants and society generally (Brown and Novak-Leonard, 2007; Lord, 2012), and a new Animating Democracy publication (Borstal and Korza, 2017) offers a way forward to value aesthetic contributions in an 'arts for change' approach.

New leadership for the National Endowment for the Arts

In 2009, President Obama appointed Rocco Landesman, a Broadway theatre owner and producer, as his new Chair of the NEA. Upon taking office, Landesman faced a pitifully small budget, a prickly Congressional climate, and a cultural world dominated by large institutions and philanthropies. Landesman, owner of four theatres on Broadway and producer of path-breaking productions such as *The Producers* and *Angels in America*, the first Broadway drama dealing with the AIDS crisis, brought deep experience with commercial arts practices to the role. Moreover his deputy, Joan Shigekawa, the long-time

Rockefeller Foundation[5] Arts Program Director, was known as a silo-buster, reaching out to non-arts colleagues, embracing a more elastic notion of 'the arts,' and persisting in innovative arts grants despite push-back from traditional arts institutions. Shigekawa introduced Chair Landesman to academic research demonstrating that neighborhood-embedded arts produce positive results for communities' jobs and business income, quality of life, public safety, and diversity (Stern and Seifert, 1998, 2007, 2008). Landesman and Shigekawa called their approach 'creative placemaking.' Landesman believed the arts in America were broadly viewed as elitist and left-wing. He coined a new slogan, 'Art Works,' which had triple meanings: the making of art (plays, books, paintings, compositions); how art works to move people; and the contributions of the arts to economic and social life. Together, they strategized how the arts could help America emerge from the throes of the recent Great Recession. Why not ask arts organizations and artists to walk out of their doors and studios and use their creative skills to work with partners to reinvigorate and help rebuild neighborhoods and communities? Why not tailor NEA funding to such partnerships?

Naming their initiative *Our Town*, they invited us to write a research-based White Paper to introduce it to their constituencies, Congress, and the public. Given only six to seven months to complete the study, we scanned the nation for mature cases of what could be dubbed 'creative placemaking,' analyzing characteristics and challenges. Senior Deputy Chairman Shigekawa asked that our case studies reflect the many dimensions of place diversity: older industrial cities, youthful suburban-style cities, small towns and ethnically distinctive places. And that diverse art forms (music, visual art, performance, design) be represented across the case studies. She also sought cases with cross-sector partnerships: at a minimum, one public and one non-profit organization. From these, we designed an analytical framework for understanding the creative placemaking process and its challenges. Chair Landesman requested that we estimate the national contribution of the arts to the national economy in our White Paper. This was daunting, given that the national income and product accounts don't have a category dubbed 'arts and culture.' We used prior research we'd conducted exploring artistic occupational concentrations in industrial sectors to roughly estimate the arts' shares of both GDP and net exports (Markusen and Gadwa Nicodemus, 2013). Subsequently, the Research Director of the NEA, Sunil Iyengar, worked with the Bureau of Economic Analysis[6] to produce annual cultural industry series: for example, in 2012, the arts generated $1.1 trillion in GDP and 4.7 million jobs, more than construction or agriculture (National Endowment for the Arts, 2013).

By the time the Markusen and Gadwa Nicodemus' study, *Creative Placemaking* (2010), was unveiled, Landesman and Shigekawa had put in place the design elements for the *Our Town* funding initiative. They required that every application must be submitted by a partnership that includes at least one public-sector agency and one non-profit, one of which must be an arts organization. They were clear that this didn't mean simply a mayor's or arts leader's sign-off on a proposal, but major buy-in and a firm commitment of money and staff time from all partnering organizations.

Multiplying the arts' loaves and fishes

In a complementary form of market research, Landesman, from the day of his arrival at the NEA, began touring the country to test the waters for *Our Town*. Alerting the press in advance and flanked by local arts leaders, Landesman pitched the stellar contributions of art and culture to mayors, journalists, and the public. In turn he listened to those he engaged with, bringing intelligence back to his team's strategic conversations. Throughout his chairmanship, he travelled around the country at least two of every four weeks as he sought to take the pulse of the nation's diverse communities, from major cities to tiny hamlets, gathering intelligence that would help the NEA to fashion its funding criteria and continually redesign the initiative.

Immediately following his move to Washington, Landesman began to leverage his tiny budget. 'Willie Sutton robbed banks because that's where the money was,' Landesman was fond of saying. 'Money, in our case, was to be found across other federal agencies and in the private sector.' Harnessing his producer talents and regarding himself informally as of Cabinet-level stature, he began dining with the Secretaries of the large, wealthy agencies – Housing and Urban Development (HUD), the Department of Transportation (DOT), and those governing health and human services, labor, environmental protection, and national defense – pitching to them how arts and culture could help them with their missions. 'The talking point stopped being "what you can do for arts," to "what the arts can do for you,"' explained Landesman in a 2015 interview with us. Rather than asking HUD or DOT for more money for the arts, he'd offer 'let us help you build more attractive light-rail stations and more vibrant communities.' Several large agencies began to incorporate the arts into major grant programs. With the US armed forces, the NEA pioneered creative arts therapy for active-duty military suffering traumatic brain injuries and post-traumatic distress.

To expand the NEA's effectiveness, Landesman and Shigekawa approached the presidents of major American foundations to create a separate and complementary funding stream. The Ford Foundation convened its counterparts to underwrite ArtPlace, an unprecedented ten-year philanthropic consortium to fund similar experiments around the country without the constraints faced by a public-sector funder. The consortium leaders recruited big banks as development loan partners and federal agencies as strategic advisors. Since their inception, *Our Town* and ArtPlace have harnessed the power of the arts, from beauty to cultural bridging, to animate community and economic development. In Detroit, they've helped local artists and entrepreneurs use pop-up spaces to re-spark the city's Avenue of Fashion.[7] In Saint Paul, artists partnered with local businesses to survive a disruptive light-rail construction project.[8] Fargo residents transformed an 18-acre barren storm-water collection basin into a community garden space that celebrates immigrant and Native American cultures.[9] In Prattsville, New

York, a local arts center led the reimagining and rebuilding of downtown following Hurricane Irene.[10] Eight years in, what have been the major challenges?

Challenges for creative placemaking: Diversity and displacement

We might have foreseen more clearly the design flaws and challenges to creative placemaking in our initial study. They were not long in emerging as funding and implementation began to spread across the country. First, the lack of racial and class diversity in the leadership of and constituencies for funded initiatives became an issue in many places, as we learned on the public-speaking and conference circuit. Second, the potential for creative placemaking to displace lower-income residents and businesses quickly became apparent and controversial. Third, and the subject of the following section, the absence of clear outcome expectations and performance standards for creative placemaking projects has prompted debate over evaluation, which is particularly significant for the NEA.

Diversity

Although some of our fourteen brief *Creative Placemaking* case studies and photos (Markusen and Gadwa, 2010, pp.31–61) underscored diversity of participation and constituencies served, none of the cases were initiated by people of color. Yet for decades, many outstanding examples of US place-embedded initiatives have been spearheaded by African Americans, Latinos, Native Americans, and Asian Americans. Though they may not have had placemaking as their primary objective, these inititatives functioned as such. Lipsitz, in *How Racism Takes Place* (2015), devotes a chapter to African-American Horace Tapscott's creation of the *World Stage* in Los Angeles and another to Rick Lowe's *Project Row Houses* initiative in Houston.[11] In retrospect, the partnership aspirations of the NEA leadership – their requirement that each prospective *Our Town* project be led by at least one public-sector and one non-profit agency – made it difficult, in an era where few city governments had much of an arts agenda at all, for us to identify completed cases led by artists of color. The NEA's focus on projects rather than organizations made it more difficult to identify initiatives by people of color as well. For so many decades, American people of color and immigrants have created their own organizations and spaces with no support from their local public sector and very little from philanthropic agencies. Worse, they sometimes encountered disparagement and open antagonism. In retrospect, had we had more time and been able to dig deeper, we would have learned about the persistent racism on the part of government agencies, officials, and funders that make it so difficult for arts organizations of color to thrive in the places they wish to be located.

For instance, we considered the case of MACLA (*Movimiento de Arte y Cultura Latino Americana*)[12] in San Jose, California, founded in 1989 to provide arts programming as a vehicle for civic dialogue and social equity. MACLA has served as multi-disciplinary arts center and anchor for its neighborhood and the city- and county-wide Latino community for a quarter-century. Recently MACLA undertook a community-development project in its neighborhood where immigrant women could learn silk screening and help mobilize neighborhood and economic development projects (Alvarez, 2005; Matthews, 2005). In 2009, when we surveyed possible case studies, MACLA was engaged in principled disagreements with the city's economic development agencies over its location. From the start, MACLA's founding Director, Maribel Alvarez, and her Board had committed to a downtown location and aspired to own its own building. The City of San Jose had other ideas: it was investing in its Mexican Heritage Plaza elsewhere in the city and pressured MACLA leaders to move there. Alvarez describes how the city pitted one ethnic group against another and contended: 'Isn't a Mexican Center enough?' No, responded the youthful MACLA: 'We are not just folkloric – we want to be part of contemporary cultural conversations! We want to nurture all of Latin American arts and cultures, and we want to stay downtown!' Current Director Angie Helstrup-Alvarez recounts MACLA's long struggle to buy its building. In the late 1990s, the San Jose Redevelopment Agency bought and land-banked it, a de facto excuse for not improving the space for existing tenants. Just before the 2008 economic downturn, the Agency planned to convert it to pricey condos, a plan deferred during the Great Recession. Contesting a new condo plan in 2009, MACLA's Board and Director asked the Agency if they couldn't buy the building themselves. They succeeded because some key external stakeholders, through Leveraging Investments in Creativity (LINC) and eventually the Ford Foundation, came through with financing. After many setbacks and delays, MACLA closed escrow in May of 2013. In 2017, Helstrup-Alvarez affirmed in a recent phone conversation: 'If we hadn't bought the building, there would be no Mexican-American cultural presence in downtown San Jose.' MACLA's leaders speak of the pervasive racism that they and other arts organizations of culture have encountered over the decades, manifest in researchers' recent empirical and qualitative accounts as well (Sidford, 2011; Markusen et al., 2011). The California arts-funding foundations did not regard MACLA and others as competent, especially because they did not have CEOs or CFOs on their Boards, namely people who could write those checks themselves. 'If your Board members aren't hanging out at the Aspen Institute, you aren't apt to be on the philanthropic radar,' says Helstrup-Alvarez. Though increasing numbers of US arts organizations are now hiring people of color, few of them are leading organizations devoted to people of color (Schumacher et al., 2016), and instead frame their different approaches along a 'cultural equity' continuum from 'diversity' to 'self-determination.' Maribel Alvarez, also in a recent phone conversation, reflected that today, MACLA is one of the very few cultural organizations of color in the US to own valuable downtown real estate. Both MACLA leaders credited new initiatives at the time like Miguel Garcia's Ford Foundation initiative *Shifting Sands*, LINC, and Project for Livable Communities with developing peer-group networks among place-embedded arts organizations of color.

Despite challenges in finding funding and public-sector partners, many African-American-led arts and cultural organizations have led in placekeeping, a practice Bedoya (2013, 2014) counterposes to placemaking. African-American churches across the US have long served as venues for the teaching and performance of music, movement, and visual art. Despite the shackles of Jim Crow laws, they have stabilized and deepened the cultural vitality of entire neighborhoods and rural communities. Since the early 1990s, the work of Ashe Cultural Center,[13] founded by writer/producer Carol Bebelle and artist Douglas Redd and closely allied with Black churches, has consistently engaged in placekeeping in its immediate Central City neighborhood and for New Orleans citywide. Its artistic missions center on diversity and place – 'to exhibit a more positive and inspiring view of African Americans for New Orleans' – and 'to foster and participate in community organizing and networking efforts in Central City to improve the opportunity to connect with and access mainstream opportunities' (Ashe Cultural Arts Center, 2017).

Young new organizations of color often concentrate on direct art and culture services for community members. Sometimes this mission is strongly place-focused. In 1996, three young African Americans with roots in the poverty-stricken Northwest Minneapolis area formed Juxtaposition[14] to teach youth from the area art and encourage them to become artists, particularly focussing on employment/career paths. Blending urban art and fine art, they use public art as a catalyst for broader community dialogue and a vehicle for verbalizing ideas about who people are and what community means. They work to transform their campus and neighborhood physically too. Since 2004, Juxtaposition has actively used its presence and programming to transform its West Broadway home, once a seedy-looking commercial strip populated by liquor stores, fast food outlets, and freeway exchanges, into an attractive and welcoming community center. Its decades-long project, *Remix: Creating Places for People on West Broadway*, a collaboration with the local community development group and university, engages area youth in creating safe spaces using art (Markusen and Johnson, 2006, pp.78–83; Gadwa Nicodemus et al., 2017).

Scholarship and literature addressing the origins of and contemporary scourge of racism in American communities illuminates the way that American political systems (e.g. the American constitution, Congress, the Supreme Court, Reconstruction, Jim Crow, immigration laws and practices, land distribution, and military policies, including many twentieth-century and contemporary practices) have operated to create and sustain racist practices (Feagin, 2013; Lipsitz, 2015). Cultural segregation is a major culprit in this drama. Reading Alice Walker's *In Search of our Mothers' Garden: Womanist Prose* (1983) and Ta-Nehisi Coates' *Between the World and Me*, especially the chapter on his first year at Howard University, reveals how Americans have been deprived of exposure to the richness of African-American (and African) literature and culture more broadly. Many writers and artists have similarly documented the exclusion of cultural expression and contributions by Native Americans, Asian Americans, and Latinos. This continues in many places today. Carolina Sarmiento's (2014) dissertation on the creation of a Mexicano immigrant arts organization and center in Santa Ana, California, documents how this cultural group faced stiff resistance from and disparagement by the city's all-Latino city council, whose downtown development initiatives intentionally and continually are displacing Mexicano working class cultural spaces and businesses. Thus immigrant status and class also affect creative placemaking conflicts and outcomes. These reflections on diversity suggest that creative placemaking must attend to the issues of race, ethnicity, LGBTQI+, class, and immigrant communities in the design and implementation of programs. Just as educational access, especially at the post-secondary level, has improved greatly from affirmative admissions and action programs, creative placemaking must make a priority of diversity, encouraging stated intent to affect systemic change, especially in this decade of heightened racial and immigration tensions.

Displacement

Almost immediately, the NEA and ArtPlace initiatives ran up against the thorny phenomenon of displacement. Because multiple aspirations for *Our Town* funding initiatives envisioned economic development as one of several desired outcomes, the potential for displacing low-income homeowners and renters, businesses, and jobs was almost immediately controversial. Authors researching gentrification – the movement of higher-income and wealthier households into lower-income urban neighborhoods – attribute a role in this process to artists and cultural activity as attractors (Deutsche and Ryan, 1984; Zukin, 1987). This view – and the lack of solid empirical evidence for it – has been questioned by many, including a brilliant essay by art critic Davis (2013), a thought piece by Gadwa Nicodemus (2013a), and an evidence-based rebuttal in Markusen (2006). Elsewhere, Bedoya and Markusen argue that displacement rather than gentrification is a superior way of conceptualizing the equity challenges of creative placemaking (Markusen and Bedoya, 2017). Bedoya, an artist and arts administrator, made substantial contributions to the debate over creative placemaking in two articles and in public speeches in many venues across the US. In the first of these, *Placemaking and the Politics of Belonging and Dis-belonging* (2013), Bedoya argues that the concept of creative placemaking lacks an awareness of the politics of belonging and dis-belonging that operate in civil society. 'Before you have *places of belonging*,' Bedoya writes, 'you must feel you *belong*. Before there is the vibrant street one needs an understanding of the social dynamics on that street – the politics of belonging and dis-belonging at work in placemaking in civil society.' In a second contribution, *Spatial Justice: Rasquachification, Race and the City*, Bedoya (2014) explores the placekeeping practice of rasquachification, used by Latinos in claiming their neighborhoods from the 'white imaginary aesthetic' often enforced by planning practices.

Displacement is deeply embedded in American (and capitalist) law and logic. These forms of profit-taking foster a permanent built-environment industry that is powerful in achieving large government infrastructure commitments and laws that permit continued expropriation of land. These actors increasingly focused on urban land uses in the twentieth century, setting in motion a continual movement of middle-class whites out of the central city (and countryside) towards stratified suburbia, and, more recently, profiting from a reversal of these patterns through revitalization of core cities. Markusen and Bedoya (2017) place responsibility for gentrification on these forces and document the ways that communities can use various political tools to stop displacement.

Some economic development can be place-reinforcing. Some neighborhoods – especially those with empty storefronts and vacant homes – can support more housing and jobs, occurring without displacement. In a nuanced analysis of gentrification in dozens of neighborhoods in the San Francisco Bay area, Chapple and Jacobus (2009) identified cases where modest amounts (and a moderate pace of construction) of new retail and middle-class housing units have avoided displacement and improved the lives of current lower income residents – offering new jobs and better consumer opportunities without raising land values, housing prices, rents, or property taxes. In other cases, they found displacement. In another instance, a philanthropy-subsidized negotiation with national grocery chains brought large supermarkets back into Philadelphia neighborhoods and improved the quality of food offered without causing gentrification and displacement (Personal communication).

How can displacing gentrification be avoided? Bedoya and Markusen argue for vigilance in funding and program design to ensure that low-income residents and small-business owners are not forced out by funded projects. Rent control has worked in many cities to preserve affordability and prevent speculative hikes in rent and housing prices (Blumgart, 2015). Community-benefit agreements are often negotiated among citizens, developers, and local governments to shape the conditions of redevelopment, including preservation of affordable housing and provision of open space (Gross, 2005; Marcello, 2007). Community-development corporations' housing projects help curb displacement. Community land trusts offer community members a way of demarketizing housing, retail, and cultural property that eliminates ownership gains from sales and reins in gentrification pressures (Greenstein and Sungu-Eryilmaz, 2007a, 2007b). Some communities have engaged non-profit organizations, such as Artspace,[15] to build affordable artist housing and cultural creation and presentation spaces. Empirical studies by Gadwa Nicodemus (2010) and Gadwa Nicodemus and Muessig (2011) on five Artspace buildings built in the 1990s found little gentrification in surrounding neighborhoods and considerable benefits for artists, the larger arts community, local commercial businesses, and local government (see Markusen and Bedoya (2016) for sources on these tools). Above all, designers and funders of creative placemaking initiatives must require that proposed projects actively address the potential for displacement and provide alternatives and guarantees that will ensure placekeeping for current residents and businesses, especially within low-income and diverse communities.

Evaluating creative placemaking initiatives

Demonstrating successful outcomes was a goal from the start for both NEA and ArtPlace. Lead program managers scrambled to articulate and measure results, especially challenging if cultural diversity and placekeeping objectives matter. Markusen's critique (2013) helped to inform a much deeper and nuanced approach. The art of the 'creative' in creative placemaking means almost certainly that results will diverge from initial intentions. Evaluation exercises often avoid addressing explicitly the problem of conflicting values and goals among the massive development industry (often quite powerful locally) and artists, art organizations, neighbourhood leaders, and elected officials.

Efforts to define success and design indicators

Early on, both the NEA and ArtPlace attempted to design 'indicators' that could be used to gauge outcomes (National Endowment for the Arts, 2012; ArtPlace, 2012a, 2012b). Using external sources, each funder designed place-specific measures that would enable creative placemakers to compare their projects with those of others. 'Over the next year or two,' wrote Jason Schupach, the NEA's Director of Design,

> we will build out this system and publish it through a website so that anyone who wants to track a project's progress in these areas (improved local community of artists and arts organizations, increased community attachment, improved quality of life, invigorated local economies) will be able to do so, whether it is NEA-funded or not. They can simply enter the time and geography parameters relevant to their project and see for themselves.
>
> Jason Schupbach (2012)

On the speaking circuit, I (Markusen) began to field distressed questions from grantees and potential creative placemakers. In June of 2012, at the annual meetings of the Alliance for Media Arts and Culture,[16] my fellow plenary panel members, all involved in one or more ArtPlace- or *Our Town*-funded projects, expressed anxiety and confusion, baffled by the one-measure-fits-all nature of the indicators, especially in the absence of formal and case-tailored evaluation. In 2012, ArtPlace produced a set of 'vibrancy' indicators designed to capture creative placemaking success (ArtPlace, 2012a, 2012b). 'Vibrancy' cannot be precisely defined, and thus it joins a long list of impressively fuzzy concepts, a notion that I coined a decade ago to critique planners and geographers' enamoration with concepts like 'world cities' and 'flexible specialization' (Markusen, 2003). A fuzzy concept is one that means different things to different people, but flourishes precisely because of its imprecision. The 'indicator' approach left ArtPlace immediately open to trenchant critiques (Frank, 2013; Markusen, 2012, 2013). Ideally, creative placemaking evaluations will monitor the performance of each awardees' progress towards initial aspirations of its leaders and funders. Because projects should be accountable to local stakeholders as well as funders, grantees should be able to make course corrections in goals as projects unfold. Evaluation should monitor process as well as outcomes, so that other practitioners can learn from them, accelerating the learning curve.

Then there is the multi-causality challenge. Project outcomes, whether enthusiastic or disappointing, cannot be attributed solely to a particular creative placemaking intervention, another reason for being cautious about relying on area indicators. Property values may fall due to an environmentally obnoxious plant start-up or rise (and cause displacement) due to a new light-rail station. Even if indicators were correctly calibrated, many would fail because data are not available at spatial scales corresponding to creative placemaking target areas such as neighborhoods. Furthermore, many creative placemaking efforts operate at multiple scales. Gadwa Nicodemus (2010) and Gadwa Nicodemus and Muessig (2011) document how artists, non-profit arts organizations, and commercial enterprises (e.g. cafés) in artist live-work buildings variously hoped that the project would support resident artists but also have a positive impact on the regional arts community, neighborhood commercial activity, crime rates, and local property values. Gadwa Nicodemus' studies demonstrate how challenging it is to document impact along multiple dimensions, spatial scales, and time periods.

What about arts and cultural outcomes?

Ironically, early evaluation efforts included few arts and cultural dimensions. Missing were measures of success in bringing diverse people together to celebrate, inspire, and be inspired, goals stated in our *Creative Placemaking* white paper. Shouldn't creative placemaking advance the intrinsic values and impact of the arts: heighten and broaden arts participation, preserve cultural traditions, improve the quality art offerings, and produce both beauty and critical perspectives on society? Are artists and organizations, whose greatest skills emerge from the arts world, to be judged only on their impact outside of this core? Prototypes were available on this front. Under Maria Rosario Jackson's leadership, the Urban Institute completed a landmark study of cultural vitality in American communities that includes a set of arts- and culture-focused indicators (Jackson et al., 2006). One leading arts consultancy, WolfBrown, offers a website, intrinsicimpact.org, to 'change the conversation about the benefits of arts participation, disseminate up-to-date information on emerging practices in impact assessment, and encourage cultural organizations to embrace impact assessment as standard operating practice.' The research of Novak-Leonard and Brown advances our ability to measure the impact of interventions on arts participation and 'intrinsic' (pleasure, learning, emotional experience, and so on) outcomes (Brown and Novak, 2007; Brown and Novak-Leonard, 2011; Novak-Leonard and Brown, 2011; Brown, 2012). Animating Democracy's (Borstal and Korza, 2017) ambitious agenda for linking aesthetic concerns to social justice is another step forward.

The combined debate on indicators and evaluation bore fruit. Both the NEA and ArtPlace changed course as their programs rolled out, moving away from indicators as an evaluation tool. Both took the aspirations of their grantees as a starting point for evaluation, learning from the experience of each as they moved forward. The NEA has recently developed a theory of change and measurement model for *Our Town*. The evaluation tool accommodates considerable variation at the individual project level and is designed to connect practitioners to evaluation-related tools. In addition to documenting knowledge gained and impacts of its major Community Development Investment program, ArtPlace's evolving research and evaluation work features field scans that investigate how arts and cultural practices intersect with other community-development sectors (for example, housing and public safety.) Several resulting studies are models of complex and nuanced site-based evaluations (Gadwa Nicodemus et al., 2016; Lee et al., 2016; Tebes et al., 2015).

Is creative placemaking a worthy and enduring policy initiative?

The NEA *Our Town* and companion ArtPlace initiatives have significantly improved the understanding of the role of artists and arts capabilities to serve communities while making substantial contributions to demographic stabilization and cultural engagement in many places. We are struck by the range and size variation of communities across the globe who have picked up the creative placemaking moniker, convened conferences, brought it into the policy sphere with their own new and appropriate twists, and produced real progress on the local level. The recently formed Alliance for the Arts in Research Universities (a2ru)[17] consortium of university faculty across the arts and sciences has researched and continues to explore creative placemaking as a central policy initiative (Cardenas, 2016). Yet in some cases, creative placemaking initiatives may have worsened living conditions for lower income residents (often minorities and immigrants) and renters, negatively affecting the integrity of existing local cultures and undercutting the viability of community bonds. We remain hopeful that debates about diversity, displacement, and evaluation are generating positive changes in the design of the funding streams and outcome assessments. For certain, over a decade of these initiatives, creative placemaking efforts are incorporating many more people of all types in the debate over who, how and why we should support arts and cultural activity. Will creative placemaking disappear if the NEA is defunded and ArtPlace gracefully shuts its doors at the end of its ten-year tenure? We expect that it will not, but that its varieties and its protagonists have and will continue to morph. We are optimistic about its long-term contributions and hopeful for greater gains for long-discriminated against racial and immigrant groups, for greater public participation, for a flourishing of participatory arts, and as Roberto Bedoya (2013, 2014) demands, 'placekeeping,' as enduring features.

Notes

1 National Endowment for the Arts, www.arts.gov/
2 ArtPlace, www.artplaceamerica.org/
3 *Our Town*, www.arts.gov/grants-organizations/our-town/introduction
4 Americans for the Arts, www.americansforthearts.org/

 5 Rockefeller Foundation, www.rockefellerfoundatioorg/
 6 Bureau of Economic Analysis, http://bea.gov/
 7 Avenue of Fashion, www.avenueoffashion.com
 8 Irrigate, https://springboardforthearts.org/programs/irrigate/
 9 The Fargo Project, www.thefargoproject.com
10 Prattsville Art Center www.artplaceamerica.org/funded-projects/prattsville-art-center-and-residency
11 Project Row Houses, https://projectrowhouses.org
12 MACLA/Movimiento de Arte y Cultura Latino Americana, http://macla.org/
13 Ashe Cultural Arts Center, http://ashecac.org/main/
14 Juxtaposition, http://juxtapositionarts.org
15 Artspace www.artspace.org/
16 The Alliance for Media Arts and Culture, www.thealliance.media/
17 Alliance for the Arts in Research Universities, https://a2ru.org

References

Alvarez, M. (2005). *There's Nothing Informal about It: Participatory Arts within the Cultural Ecology of Silicon Valley*. San Jose, CA: Cultural Initiatives Silicon Valley. Available from: www.ci-sv.org/pdf/MAlvarez_PA_study.pdf [Accessed: 23 December 2017].

Americans for the Arts. (2016). *Arts and Economic Prosperity V*. Washington, DC: Americans for the Arts. Available from: www.artsusa.org/information_services/research/services/economic_impact/iv/reports.asp [Accessed: 23 December 2017].

ArtPlace. (2012a). *Vibrancy definitions*. Available from: www.artplaceamerica.org/loi/. [Accessed: 23 December 2017. Page no longer available.].

ArtPlace. (2012b). *Vibrancy indicators*. Available from: www.artplaceamerica.org/articles/vibrancy-indicators/. [Accessed: 23 December 2017. Page no longer available.].

Ashe Cultural Arts Center. (2017). *History and values*. Available from: www.ashecac.org/main/index.php/about-us/history.html. [Accessed: 23 December 2017].

Bedoya, R. (2013). 'Placemaking and the politics of belonging and dis-belonging, *Grantmakers in the Arts Reader*, 24(1). Available from: www.giarts.org/article/placemaking-and-politics-belonging-and-dis-belonging. [Accessed: 23 December 2017].

Bedoya, R. (2014). 'Spatial justice: Rasquachification, race and the city', *Creativetime Reports*. Available from: http://creativetimereports.org/2014/09/15/spatial-justice-rasquachification-race-and-the-city/. [Accessed: 23rd December 2017].

Blumgart, J. (2015). 'In defense of rent control', *Pacific Standard*, April 1. Available from: www.psmag.com/business-economics/in-defense-of-rent-control. [Accessed: 23 December 2017].

Borstal, J. and Korza, P. (2017). *Aesthetic Perspectives: Attributes of Excellence in Arts for Change*. Washington, DC: Animating Democracy. Available from: www.animatingdemocracy.org/aesthetic-perspectives [Accessed: 23 December 2017].

Brown, A. (2012). 'All the world's a stage: Venues and settings, and the role they play in shaping patterns of arts participation',*GIA Reader*, 23(2), Summer. Available from: www.giarts.org/article/all-worlds-stage. [Accessed: 23 December 2017].

Brown, A. and Novak, J. (2007). *Assessing the Intrinsic Benefits of a Live Performance*. WolfBrown. January. Available from: http://wolfbrown.com/images/books/ImpactStudyFinalVersionFullReport.pdf [Accessed: 23 December 2017].

Brown, A. and Novak-Leonard, J. (2011). *Getting in on the Act: How Arts Groups are Creating Opportunities for Active Participation.* San Francisco: The James Irvine Foundation. Available from: http://irvine.org/images/stories/pdf/grantmaking/Getting-in-on-the-act-2011OCT19.pdf [Accessed: 23 December 2017].

Cardenas, E. (2016). 'A snapshot of creative placemaking in higher education' [online] in *Alliance for the Arts in Academia.* Available from: http://a2ru.org/. [Accessed: 23 December 2017].

Chapple, K. and Jacobus, R. (2009). 'Retail trade as a route to neighborhood revitalization', in Pindus, N., Wial, H., Wolman, H., and Bowen, J. (eds.), *Urban and Regional Policy and Its Effects, Volume II.* Washington, DC: Brookings Institution-Urban Institute.

Davis, B. (2013). 'Are artists to blame for gentrification? Or would SoHo, Chelsea and Williamsburg have gentrified without them?', *Slate*, October 15. Available from: www.slate.com/articles/life/culturebox/2013/10/are_artists_to_blame_for_gentrification_or_would_soho_chelsea_and_bushwick.html. [Accessed: 23 December 2017].

Deutsche, R. and Ryan, C. G. (1984). 'The fine art of gentrification, *October*, 31.

Feagin, J. (2013). *The White Racial Frame* (2nd ed). New York: Taylor and Francis.

Florida, R. (2002). *The Rise of the Creative Class.* New York: Basic Books.

Frank, T. (2013). 'Dead end on shakin' street, *The Baffler*, 22. Available from: www.thebaffler.com/past/dead_end_on_shakin_street. [Accessed: 23 December 2017].

Gadwa, A. (2010). *How Artist Space Matters: Impacts and Insights from Artspace Project's Developments.* Minneapolis:. Metris Arts Consulting.

Gadwa, A. and Muessig, A. (2011). *How Artist Space Matters II: The Riverside, Tashiro Kaplan and Insights from Five Artspace Case Studies and Four Cities.* Minneapolis: Artspace Projects and Metris Arts Consulting.

Gadwa Nicodemus, A. (2013b). 'Fuzzy vibrancy: Creative placemaking as ascendant U. S. cultural policy', *Cultural Trends*, 3–4.

Gadwa Nicodemus, A., Engh, R., and Walker, C. (2017). *Not Just Murals: Artists as Leaders in Community Development.* Local Initiatives Support Corporation, Fall 2017.

Greenstein, R. and Sungu-Eryilmaz, Y. (2007a). 'Community land trusts: Solution for permanently affordable housing', *Land Lines Magazine*, January: 8–13. Available from: www.lincolninst.edu/search/site/Greenstein?f%5B0%5D=mediatype_levelone%3A5. [Accessed: 23 December 2017].

Greenstein, R. and Sungu-Eryilmaz, Y. (2007b). *A National Study of Community Land Trusts.* Cambridge, MA: Lincoln Institute of Land Policy. July. Available from: www.lincolninst.edu/search/site/Greenstein?f%5B0%5D=mediatype_levelone%3A5 [Accessed: 23 December 2017].

Gross, J. (2005). *Community Benefits Agreements: Making Development Projects Accountable.* Washington, DC: Good Jobs First and the Partnership for Working Families. Available from: www.goodjobsfirst.org/sites/default/files/docs/pdf/cba2005final.pdf [Accessed: 23 December 2017].

Ivey, B. (2008). *Arts, Inc: How Greed and Neglect Have Destroyed Our Cultural Rights.* Berkeley: University of California Press.

Jackson, M. R., Kabwasa-Green, F. and Herranz, J.. (2006). *Cultural Vitality in Communities: Interpretation and Indicators.* Washington, DC: The Urban Institute. December. Available from: www.urban.org/url.cfm?ID=311392 [Accessed: 23 December 2017].

Killacky, J. (2011). 'Regrets of a former arts funder'; in *Blue Avocado*, June 23. Available from: http://blueavocado.org/node/664. [Accessed: 23 December 2017].

Killacky, J. (2014). 'Blood sacrifice' in *American Theatre Magazine*, January. Available from: www.flynncenter.org/blog/2014/01/blood-sacrifice/. [Accessed: 23 December 2017].

Kreidler, J. (1996). 'Leverage lost: The nonprofit arts in the post-ford era' in *In Motion Magazine*. Available from: www.inmotionmagazine.com/lost.html. [Accessed: 23 December 2017].

Lee, S., Linett, P., Baltazar, N., and Woronkowicz, J. (2016). *Setting the Stage for Community Change: Reflecting on Creative Placemaking Outcomes*. Mortimer & Mimi Levitt Foundation. Available from: http://levitt.org/research. [Accessed: 23 December 2017].

Lipsitz, G. (2015). *How Racism Takes Place*. Philadelphia: Temple University Press.

Lord, C. (ed). (2012). *Counting New Beans: Intrinsic Impact and the Value of Art*. San Francisco: Theatre Bay Area.

Marcello, D. (2007). 'Community benefit agreements: New vehicle for investment in America's neighborhoods', *Urban Lawyer* 39(3).

Markusen, A. (2003). 'Fuzzy concepts, scanty evidence, policy distance: The case for rigor and policy relevance in critical regional studies', *Regional Studies* 37(6/7).

Markusen, A. (2006). 'Urban development and the politics of a creative class: Evidence from the study of artists', *Environment and Planning A* 38(10).

Markusen, A. (2012). 'Why creative placemaking indicators won't track creative placemaking success', *Createquity*, November 9. Available from: http://createquity.com/2012/11/fuzzy-concepts-proxy-data-why-indicators-wont-track-creativeplacemaking-success.html. [Accessed: 23 December 2017].

Markusen, A. (2013). 'Fuzzy concepts, proxy data: Why indicators won't track creative placemaking success', *International Journal of Urban Sciences* 17(3).

Markusen, A. (2014). 'Creative placemaking: Partnering with arts and culture to animate cities', *Pilsen 2015*. Prague: Aspen Institute, December.

Markusen, A. and Bedoya, R. (2017). *Race, and Placekeeping: A Reframing of the Gentrification Debate*, Working paper, Project on Regional and Industrial Economics, University of Minnesota.

Markusen, A. and Gadwa, A. (2010). *Creative Placemaking*. Washington, DC: Mayors' Institute on City Design and the National Endowment for the Arts. October. Available from: www.arts.gov/publications/creative-placemaking [Accessed: 23 December 2017].

Markusen, A. and Gadwa Nicodemus, A. (2013). 'Spatial divisions of labor: How key worker profiles vary for the same industry in different regions', in McCann, P., Hewings, G., and Giarattani, F. (eds.), *Handbook of Economic Geography and Industry Studies*. London: Edward Elgar.

Markusen, A. and Gadwa Nicodemus, A. (2015). 'City creative industry strategies: Unique American cases', *Aspen Review Central Europe*, 4.

Markusen, A. and Johnson, A. (2006). *Artists' Centers: Evolution and Impact on Careers, Neighborhoods and Economies*. Minneapolis: Humphrey Institute of Public Affairs, University of Minnesota, February. Available from: www.hhh.umn.edu/projects/prie/PRIE–publications.html. [Accessed: 23 December 2017].

Markusen, A., Gadwa, A., Barbour, E., and Beyers, W. (2011). *California's Arts and Cultural Ecology*. San Francisco, CA: The James Irvine Foundation, September.

Markusen, A., Wassall, G., DeNatale, D., and Cohen, R. (2008). 'Defining the creative economy: Industry and occupational approaches',*Economic Development Quarterly* 22(9).

Matthews, L. (2005). 'The ties that bind: A community-based art project in Silicon Valley', in Atlas, C., and Korza, P. (eds.), *Critical Perspectives: Writing on Art and Civic Dialogue*. Washington, DC: Americans for the Arts.

McCarthy, K., Heneghan Ondaatje, E., Zakaras, L., and Brooks, A. (2004). *Gifts of the Muse: Reframing the Debate about the Benefits of the Arts*. Santa Monica, CA: RAND

Corporation. Available from: www.rand.org/pubs/monographs/MG218 [Accessed: 23 December 2017].

National Endowment for the Arts. (2012). *Creative placemaking*. November 13. Available from: www.fbo.gov/?s=opportunity&mode=form&id=3e198017f18bfd723 f557702a7b46bca&tab=core&_cview=1. [Accessed: 23 December 2017].

National Endowment for the Arts. (2013). NEA guide to the US arts and cultural production satellite account. www.arts.gov/publications/nea-guide-us-arts-and-cultural-production-satellite-account. [Accessed: 23 December 2017].

Novak-Leonard, J. and Brown, A. (2011). *Beyond Attendance: A Multi-Modal Understanding of Arts Participation*. Washington, DC: National Endowment for the Arts, February.

Sarmiento, C. (2014). *Strategies of Fire and ICE: Immigration, Cultural Planning and Resistance*. Santa Ana: University of California Irvine doctoral dissertation.

Schumacher, C., Ingersoll, K., Nzinga, F., and Moss, I. D. (2016). 'Making sense of cultural equity: When visions of a better future diverge, how do we choose a path forward?' in *Createquity*. Available from: http://createquity.com/2016/08/making-sense-of-cultural-equity/. [Accessed: 23 December 2017].

Schupbach, J. (2012). *Creative Placemaking—Two Years and Counting!* Washington, DC: National Endowment for the Arts. Available from: http://artworks.arts.gov/?p=13382 [Accessed: 23 December 2017].

Sidford, H. (2011). *Fusing Arts, Culture and Social Change: High Impact Strategies for Philanthropy*. Washington, DC: National Center for Responsible Philanthropy. October. Available from: www.ncrp.org/paib/arts-culture-philanthropy [Accessed: 23 December 2017].

Stern, M. and Seifert, S. (1998). *Community Revitalization and the Arts in Philadelphia*. Philadelphia, PA: University of Pennsylvania, Social Impact of the Arts Project. Available from: www.sp2.upenn.edu/SIAP [Accessed: 23 December 2017].

Stern, M. and Seifert, S. (2007). *Cultivating 'Natural' Cultural Districts*. Philadelphia: The Reinvestment Fund. Available from: www.sp2.upenn.edu/SIAP [Accessed: 23 December 2017].

Stern, M. and Seifert, S. (2008). *From Creative Economy to Creative Society*. Philadelphia: The Reinvestment Fund. Available from: www.sp2.upenn.edu/SIAP [Accessed: 23 December 2017].

Tebes, J., Matlin, S., Hunter, B., Thompson, A., Prince, D., and Mohatt, N. (2015). *Porch Light Program: Final Evaluation Report*. New Haven: Yale University, School of Medicine. Available from: www.consultationcenter.yale.edu/798_218966_Porch_Light_Program_Final_Evaluation_Report_Yale_University.pdforconsultationcenter.yale.edu. [Accessed: 23 December 2017].

Walker, A. (1983). *In Search of Our Mothers' Gardens*. New York: Harcourt, Brace, Jovanovich.

Zukin, S. (1987). 'Gentrification: Culture and capital in the urban core', *Annual Review of Sociology*, 13.

2 Spaces of vernacular creativity reconsidered

Tim Edensor and Steve Millington

Abstract

Our edited volume, *Space of Vernacular Creativity: Rethinking the Cultural Economy* (2009) critically responded to the preoccupations that had dominated writing on placemaking in, particular culture-led regeneration. With the inception of creative placemaking and its international uptake, we revisit the book's key imperatives: to decentre instrumental and reductive conceptions about the location of creativity that were oriented around dominant notions of the 'creative economy', and interrogate how alternatively, creativity might be far more expansively conceived, within this contemporary frame. We identify what remains salient in our original arguments, which areas we ignored and underplayed, and key contributions that have advanced thinking about creative geographies within the context of creative placemaking since the book's publication.

Introducing the vernacular

In *Spaces of Vernacular Creativity* (Edensor et al., 2009), we sought to extend understandings beyond the narrow prescriptions of culture-led regeneration and the normative framework established in economic geography to identify cultural clusters (Mommaas, 2004), and explore creativity in a broader sense, focusing on the relationship between creativity and place, and how creativity could be generative of non-economic values. We wanted to provide a more constructive riposte to the many critiques of the 'creative city' and the 'creative class' script popularised by Florida (2002) and Landry (2000), to uncover practices that stood in opposition to the alignment of creativity with the cool, sophisticated and metropolitan. We asked: if there is a creative class, then who might be the uncreative classes against whom they are implicitly constructed? And if there are cool places to which this creative class gravitate, then what does this mean for all those other places that fail to make the cut on creativity indices? Instead, we wanted to embrace marginal, un-sexy and less-than glamorous unspectacular creative practices and reveal their role in enhancing the lives of people in everyday spaces, to offer a more powerful and inclusive analysis of creativity and place.

As we entered the twenty-first century, creativity occupied an elevated position in governmental policy responses to the economic and political challenges of the day. In the UK, for instance, 'Cool Britannia' captured a sense of optimism that somehow, through the proliferation of arts and cultural districts, festivals and music scenes, our cities and national economy could be reinvented. Florida's contentions were conceived as providing the academic authority that supported such aims, and whatever the criticism of his key ideas and methods, we cannot ignore his influence on urban policy and civic leaders around the world in embedding the notion that the creative class can drive economic regeneration. In the time since we compiled the book, this seductive narrative has begun to fade from policy discourses following the 2008 financial crash. Priorities shifted under austerity. Arguments to maintain public subsidies for arts and culture have become marginalised in the face of calls to protect frontline services or tackle homelessness. Florida's (2017) recent tacit admission that the creative class were perhaps nothing more than this generation's gentrifiers further signifies that optimism about their central economic role is over. Nevertheless, the cultural life of places persists, by any means necessary. As arts professionals scramble for ever reduced grants, volunteers and community-led action are becoming more significant agents within the local cultural landscape. At the same time, ideas such as Lydon and Garcia's tactical urbanism (2015) have come to the fore in the absence of both public and private investment. Through a make-do attitude and experimentation, tactical placemaking interventions led by local communities have undermined the strategic thinking which underpinned the creative city script, and seem to underscore the importance of the vernacular that we highlighted.

Creative multiplicity

The first key emphasis in *Spaces of Vernacular Creativity*, as is explicit in its title, was to embark on a thorough decentring of geographies of creativity by moving towards the multiplicity of creative practices that take place elsewhere, in other sites and networks through which creative ideas and skills are produced and circulated. The overwhelming focus on the city and the city centre, implicitly conceived as the domain of the creative class, the realm in which they work, play and consume or the arts districts and cultural quarters that they fashion, has marginalised other spaces in which creative practices take place. Less glamourous parts of the city – suburbs, those areas of Victorian and Edwardian terraced housing in the UK which remain resistant to gentrification, and modernist housing schemes – were identified as less trendy or more sedate, while smaller industrial or post-industrial urban settings, market towns, villages and rural areas were largely consigned to the dominion of the irredeemably uncool and uncreative. These areas were drastically neglected while the apparently global cities of London, Berlin and New York and smaller regional cities such as Manchester and San Francisco were regarded as founts of creativity, sites where a critical mass of creatives inspired each other and consumed each other's goods and services. While galleries, coffee shops and loft studios were celebrated as sites of creativity, living rooms, sheds, garages, gardens and community centres were not. In attempting to address these geographical distortions, chapters in the book examined creative practices that take place in mundane domestic settings located in the suburbs of Toronto, the working-class estates of Sheffield, Australian country towns and community gardens in northern England.

Since then, other work has consolidated this move away from those sites exclusively identified as urban hot spots of creativity. Waitt and Gibson (2013) have detailed how unfashionable, provincial, co-operative art spaces can reinforce place-belonging, and Lobo (2017) strikingly describes how the unpromising setting of a disused underground suburban car park in Darwin, Australia, has become a convivial venue for the shared, varied creative practices of marginalised Aboriginal woman and migrants. Here, friendships have emerged through shared creative production as unexpected connections have been forged between participants within an immersive, inclusive atmosphere. More domestic spaces have been the focus of work by Platt (2017) and Hackney (2013), who consider how women's quotidian, mundane creative practices of knitting and crocheting act to craft identities and contribute to placemaking. Besides reappraising the ways in which evaluations around creativity have marginalised certain highly gendered activities, their analyses emphasise how creative production extends across the familiar realms of home and community, an arena of creativity that we somewhat neglected in *Spaces of Vernacular Creativity*. In reinforcing the creativity that inheres such domestic spaces, Lee (2010) concentrates on the more mundane practices of bed-making, cooking, tidying up and even scrubbing floors, which are also sensual, memory-filled and imaginative creative practices of homemaking. For Lee, the aesthetic effects of such daily tasks do not merely revolve around making judgements but by the 'the small pleasure afforded by the simple awareness and presence of mind during our everyday lives', and are particularly well exemplified by the traditional practice of laundry-hanging discussed by Rautio (2009), which thoroughly entangles creative imagination, memory borne of habit and multi-sensory immersion in the job at hand.

In *Spaces of Vernacular Creativity*, in exploring emergent forms of shared production across the photo-sharing Flickr network, Burgess showed how cyberspace was also a neglected realm of creativity, an arena in which connections around creative-making were vastly extended. These new media applications have subsequently multiplied with the increasing popularity of geocaching, game-playing, creative writing, video-making and the proliferation of memes. Burgess has continued to mine this area, showing that though they may be mimicked in developing commercial strategies, viral videos promote inventive adaptive engagement from a wide variety of participants, and subsequently a 'flurry of parodies, mash-ups and remixes' (2014, p.90) that exemplify the ongoing production of vernacular network creativities. Through such social networks, the boundaries between professionals and enthusiasts can become blurred, with some non-professional participants becoming widely renowned. Similarly, Vásquez and Creel (2017) highlight the imaginative 'chats' that produce a shared conviviality amongst the online community who use Tumblr blogs. In addition, in drawing upon the phenomenon of crowdsourced art, Literat and Glăveanu (2016) exemplify how the internet has enormous potential to foster a distributed creativity grounded in collaborative communication and interaction that is available for the participation of anybody who wishes to contribute. They contend that the sheer diversity of the knowledge, skills and cultural resources of these multiple participants generates a context in which creative possibilities proliferate.

Decentring creativity

Despite the abundant evidence that creative practice is much more widely distributed, reductive, elitist notions of what constitutes creativity can have profound political effects upon cultural endeavours that take place away from metropolitan taste-makers. Gilmore (2013) further explores how participatory arts policy-makers and funding bodies routinely ignore local forms of creative practices that do not accord with those they esteem, and as Miles and Ebrey (2017) argue, marginalisation and lack of support from central and local governments, cultural institutions and the private sector for smaller, non-urban settings can entrench economic decline and stagnation.

In examining the geographical focus of our 2009 volume, we too were culpable of neglecting to identify and discuss creative practices in plethora of settings. Though such an exhaustive task is beyond any one text, perhaps we should have made more extensive efforts to account for non-Western forms of creativity by more extensively considering the extraordinary creative adaptations, improvisations and in fashioning urban lives and livelihoods in African, South American and Asian cities that have been explored by many writers. Indeed, as Robinson (2006, p.4) claims, an understanding of the vitality of such urban cultures without considering 'a strong sense of the creativity of cities' truncates the potential for imagining their futures. Such an exploration, amongst many other accounts, could include references to the extraordinary creation of Chandigarh's Rock Garden, now a global tourist attraction, by Nek Chand, an 'outsider artist' who assembled a range of figural sculptures from the debris of demolished buildings (Jackson, 2002); the incredibly adaptive, improvisatory extensive trading networks established by peddlers belonging to Senegal's Murid Brotherhood (Diouf and Rendall, 2000); or the hybrid architecture of Hong Kong (Abbas, 2002) and improvisational entrepreneurial tactics and fleetingly assembled structures of Lagosians (Hecker, 2010). It might also investigate the ongoing fashioning of the anti-colonial, anti-neoliberal practice of *buen vivir* (good living for all) in Latin America (Escobar, 2010); the ever-transforming streetscapes of Indian cities that are endlessly recomposed out of recycled materials and temporarily occupied by a multitude of actors (Mehrotra, 2008); the critical political performances of Lima's street comedians (Vich, 2004); and the ongoing assemblage of informal shack dwellings in Sao Paulo's informal settlement, Paraisópolis (McFarlane, 2011). On the other hand, many urban theorists from China have eagerly grasped the creative city concept and continued the normative research agenda established in the West in analysing networks and clusters of creative businesses (for example, see Cho et al., 2018).

This brings us onto a second and related key aim of *Spaces of Vernacular Creativity*, which was to interrogate how creativity might be more expansively conceived: to escape from a rather instrumental and reductive understanding that had emerged as part of discussions about the 'creative economy' and the 'cultural industries', a conception that found great favour amongst city managers and economic strategists. Our attention, therefore, was drawn towards vernacular forms of creative endeavour amongst alternative and marginal groups, as well as cultural producers who create non-economic outcomes. In moving beyond a narrow focus on taste and aesthetics, we sought to recognise how creativity may also produce social collaboration and communication.

To return to Florida, the premium placed on promoting creativity became inextricably entangled with the much-vaunted championing of what he called the 'creative class', a group conceived as essential to the regeneration of cities that had suffered significant industrial decline. These artists, gallery owners, baristas, fashion designers, advertisers and musicians were conceived as being able to reignite an economic spark by developing cultural industries and thereby attracting new inhabitants with high disposable incomes who were lured by the promise of trendy urban environments and lifestyle accoutrements. Evidently, this depiction of a particular group of people who collectively constituted a creative class was exclusive, ruling out those who pursued creativity in less circumscribed ways. Recently, the orbit of those belonging to the creative class has expanded to incorporate those from the traditional professions of barbering and bartending who, under conditions of gentrification, are being revalued in way that recognises the creative elements that they always practised (Ocejo, 2012). Nevertheless, the durable if fluid construct of the creative class continues to perpetrate a limited notion about who and what is 'creative'. Thus, the value of creativity often remains tethered to its capacity to make profits and conforms to the entrepreneurial imperatives of city managers, reinforcing the strategic role of creative industries in economic development and urban renewal. According to such a perspective, creative production and practice, especially insofar as it involves the provision of fashion, food, culture and arts, is as integral to the promotion of the city as tolerance for social and cultural diversity.

Reductive and reified conceptions also resonate in those assertions that creativity can primarily be defined in terms of aesthetic experimentation or 'innovation'; that is, in the original work of artists, poets and musicians. Our thesis in *Spaces of Vernacular Creativity* was that these debates have produced highly restricted understandings about who is creative, what can be considered as a creative product or practice and where it is that creativity takes place. We drew upon the important arguments of Hallam and Ingold (2007) to contend that creative practice can be habitual and reiterative, is necessarily adaptive and improvisational and is equally likely to be found in collective work than individual artistic innovation. Moreover, we insisted that creativity did not need to be identified by experts – including the growing host of cultural intermediaries and media presenters – and in fact where this took place, was principally illustrative of the habitus and dispositions of those who espoused such particular tastes. These arguments have been subsequently reinforced by alternative, much more inclusive definitions such as that articulated by Waitt and Gibson, who maintain that creativity is 'a field of choices and possibilities that are set up in the tensions between being and becoming' (2013, p.75). Such tensions proliferate through social life.

One central argument in the book focused on foregrounding the qualities of vernacular and everyday creativity. All too often, such endeavours are maligned by arguments based on class-oriented values that champion certain forms of creative production in contradistinction to those that are not judged to be 'cool', sophisticated or fashionable. By contrast, we wanted to honour the non-economic values and outcomes produced by alternative, quotidian, diverse and more socially inclusive creative practices, and in this spirit, chapters discussed the Elvis festival situated in a small, rural Australian town (Gibson et al., 2009), the amusing and diverse uses of garden gnomes (Potts, 2009) and the seasonal practice of garbing the outside of houses with Christmas lights (Edensor and Millington, 2009). Such practices were community-oriented and emerged in mundane settings, and were certainly not examples of top-down urban provision.

Happily, a plethora of academic accounts about other mundane, vernacular and everyday skilled practices have emerged since the book was published. This has coincided with an upsurge in work on geographies of making and crafting (Hawkins and Price, 2018) that also honours both artistic endeavour and more vernacular, collective practices. Such accounts have foregrounded the creativity and communality that inheres in the everyday practices of hairdressing (Holmes, 2015), knitting (Price, 2015) and customising cars (Warren and Gibson, 2011), as well as the craft expressed in dry stone-walling the making and curation of neon signs (DeLyser and Greenstein, 2018), the production of hand-made surfboards (Gibson and Warren, 2014) and festive lantern-making (Edensor, 2018), creative practices that move between highly skilled work to hobbyist enthusiasms.

This inclusive shift to considering creative practices of making is also augmented in recent times by an expanded understanding of the creative economy that incorporates a range of labour practices that has surfaced through what Carr and Gibson (2016, p.299) describe as 'a renaissance in small-scale making' that has especially emerged in industrial cities, in which 're-connections are being forged with themes such as quality, providence, craft, ethics, tacit design knowledge, haptic skill and the value of physical labour'. Here then, and following Hallam and Ingold's contentions, creativity was always already embedded in all forms of industrial labour; indeed, production was dependent upon the capacities of workers to adapt and acquire a sensuous knowledge of the products that they helped to forge and assemble. As these manufacturing processes have become less familiar in many deindustrialised urban settings, the skill and know-how required to make things is being revalued, tinged with a nostalgic sense of loss. Yet though the value of making has been substantively reappraised in recent times, in certain contexts, tendencies to perpetrate hoary distinctions between crafted products and practices of making remain. For instance, away from the contexts of urban industry, in exploring the reconfiguration of rural creativity, Bell (2015) discusses how cultural intermediaries express aesthetic judgements that strongly prioritise the innovative and cutting edge over 'traditional' and customary craft items. Moreover, the primary utility of the crafts as an economic resource rather than as valuable in achieving broader social goals has been integral to recent neo-liberal British governmental agendas (Jakob and Thomas, 2017).

Similar kinds of negative appraisal also continue to surround creative production in other settings that are distant from metropolitan fashions. One example is in Blackpool, Britain's most popular holiday resort. Situated on the Lancashire coast and founded in the nineteenth century to serve as a site for pleasure and leisure for workers in the industrial urban centres of Lancashire and Greater Manchester, the town has long been associated with cheap, popular and working-class attractions, often caricatured as 'vulgar', 'tacky' and tasteless by supposedly more sophisticated cultural commentators. The town's illuminations, arranged along six miles of seafront, have attracted millions of visitors for over a hundred years, and yet are rarely subject to media reviews. They epitomise a local expression of vernacular creativity in a style that remains immune from wider fashions and notions of 'good taste' and coolness, sitting outside the professional and sophisticated circuits that have driven the accelerating internationalisation of metropolitan light festivals. Rather, the design of the illuminations – which are stand-alone installations, themed sections arranged between lampposts on either side of the seafront road, or large tableaux – follows a distinctly place-based vernacular based on longstanding aesthetics and craft know-how, and is designed to satisfy the desires of visitors to experience nostalgia, conviviality and jollity (Edensor and Millington, 2013). Situated in the resort's own illuminations depot, local designers and technicians produce and stage the annual two-month extravaganza, oblivious to the tastes of metropolitan taste-makers who are unable to recognise a creative practice that is grounded in a distinctively local place identity and history and is not concerned with posing as fashionable and cool.

Creative citizenship

As we have emphasised, the notion of the creative economy is tied to a reductive understanding that creativity is an economic resource that can be deployed to advance urban growth. In *Spaces of Vernacular Creativity* we challenged this instrumental, neo-liberal conception by foregrounding the more-than-economic forms of creativity that persist and the important social and cultural functions that they advance. Creativity, we argued, does not need to be connected with economic growth at all, but can be underpinned by other values such as generosity, care and reciprocity. As Waitt and Gibson claim, 'capitalist means of production are only one possible way of organising resource use and exchange, opening up possibilities to explore non-capitalist, anti-capitalist, non-profit, collective, informal and socialist means of production' (2013, p.77). This articulates how notions of productivity need to be defined beyond a narrow economic context, wherein the production of friendship, well-being and conviviality might be more extensively honoured as some of the consequences of creative practices, along with more tangible outcomes such as the strengthening of individual and collective capacities, community building and placemaking. Hargreaves and Hartley (2016) declare that such non-economic outcomes might constitute the more inclusive notion of 'creative citizenship', where the potential to engage in forms of creative practice can produce convivial engagements with others that generate enduring social connections. Indeed, Platt (2017) demonstrates that women's shared everyday crafting practices produce exactly these kinds of lasting friendships as well as a sense of well-being amongst participants.

In further considering more-than-economic motivations for creative practices, we can identify a range of practices that are entangled with particular political objectives and with experimenting with alternative lifestyles that move away from consumerism and towards social and environmental sustainability. As Harriet Hawkins claims, the aforementioned renewal of craft practices – or 'craftivism' – is frequently oriented around an avowedly political project that foregrounds an ethos of recycling, making-do and mending. Gregson et al. (2010) have discussed the extraordinary adaptations of materials wrenched from huge obsolete ships by Bangladeshi furniture makers and other craft workers for economic gain, innovatively refashioning industrially produced materials as part of a recycling practice that leaves little vestige of these giant vessels. At a much smaller-scale, the mundane but highly skilled practices of the host of restorers featured by Bond et al. (2013) who repair clocks, bicycles, ceramic items, books, footwear, furniture and musical instrument, amongst other objects, are the remaining exponents of a world in which repair and maintenance were once a far more commonplace activity, both as household chores and paid work. And yet it seems that such practices are once more becoming championed in the context of finite resources and disposability, as the ethical implications of revaluing these practices become more apparent. Here, practices through which objects and materials are assembled and reassembled, customised, adapted, altered and restored are generating new forms of creative know-how as well as recovering lost skills.

This evolving experimental disposition towards devising alternative everyday urban lifestyles is explored by Wendler (2016) who investigates how particular groups try-out more sustainable, equitable and collective ways of living in autonomous spaces temporarily carved out in the city. Such open-ended, improvisational spaces foster new imaginaries, propose different economies and develop distinct social relations in envisaging alternative futures. These experimental efforts to achieve more sustainable living practices are more radically exemplified by Canadian off-gridders who improvise ways of living in isolated situations, working out how to build houses and manage energy supply by practising forms of 'modest creativity and mundane intuition'. Such adaptive exercises emerge out of unfolding experiences of dwelling that deepen a relationship with place (Vannini and Taggart, 2014, p.282).

At the same time, and has so often been the case, certain non-economically motivated creative practices that purport to challenge over-commodified and over-regulated urban environments, notably those that gather under the category of tactical urbanism, have, according to Oli Mould (2014), proved susceptible to incorporation by the very agencies that they oppose. For certain urban policy strategies, rather than seeing such creative re-appropriations of urban space as illicit, have construed them as signifying coolness, as vital elements that demonstrate an area's 'vibrant' street culture. These guerrilla, participatory and pleasurable initiatives, include flash-mobbing, yarn-bombing, pop-up shops and guerrilla gardening. By being absorbed into urban marketing strategies in this fashion, such insurgent tactics are in danger of losing their critical potential. According to Mould, they become expected ingredients of the creative city rather than intrusive and disruptive to official governance and advertising, though perhaps he pessimistically overstates the extent to which all resistant practices can be incorporated in this way. As McLean (2017) exemplifies, creative expressions of resistance and solidarity can re-emerge despite the absorption of radical creative groups into local state-led, instrumental economic strategies that aim to market their 'edginess'.

Although this suggests that the economic is frequently able to co-opt forms of resistant creativity into economic and managerial strategies, the relationships between the economic and the non-economic, and between artistic and vernacular forms of creativity, are often more complex and ambiguous than we suggested in *Spaces of Vernacular Creativity*. To illustrate these complexities, we draw on the example of the Moonraking Festival held bi-annually in the small West Yorkshire mill town on Slaithwaite (Edensor, 2018). Thirty years since its inception, the event remains popular amongst the townsfolk. It is based on a local myth from the early eighteenth century based on the illegal trade in alcohol, supplied covertly by barges on the local canal into which barrels were dropped and retrieved by men from the town. Upon being apprehended by vigilant customs officers, the men pretended to be drunk, and in that state declared that they were trying to fish the moon out of the canal, thereby duping the officers. This incident of local cunning forms the basis for a lantern parade that commences at the canal basin, at which a large paper lantern in the shape of a moon is hoisted to the front of the procession, which subsequently makes its way around the village. Hundreds of the townsfolk participate in carrying illuminated lanterns fashioned in accordance with the theme that has been chosen that year. The biannual recurrence of the festival has generated the development of considerable local skill in lanternmaking, through the organisation of workshops. Several local participants who have been steeped in the festival since childhood have become professional lantern-makers, selling their expertise in workshops organised in preparation for this festival and other events, and making high-quality installations and lanterns for festive display. Here, communal, non-economic creative practice has eventuated in providing some contributors with a livelihood as makers, artists and designers.

Any reductive attempts to circumscribe the geography of creativity and delineate the kinds of activities and people that should be associated with creativity, are, we believe, doomed to failure. Creativity proliferates and seethes in everyday life and in quotidian spaces; it is present in the most mundane domestic practices, in work procedures and leisure activities. It is not merely expressive of a unique individual aptitude, but can be shared and produced in convivial settings, it may reside in experimental or reiterative approaches to living, making and socialising. It most certainly cannot only be associated with entrepreneurs and artists, and is undoubtedly located in settings that are far from urban centres. This greater inclusivity is being borne out by the incorporation of a host of domestic, craft, industrial, artistic, communal and traditional practices into accounts that are widening the scope of what we might consider creative. We welcome the rejection of creativity as intrinsically economic, urban and singularly individualistic, claims that have been greatly expanded by scholars since the publication of *Spaces of Vernacular Creativity*. We anticipate that elitist, class-ridden definitions will be more widely rejected, recognised as signifying banal efforts to acquire cultural capital and status, and the protean nature of creativity will become ever more apparent.

Acknowledgements

Thanks to Cara Courage and Anita McKeown for their supportive, critical and patient support, and to Talia Melic for reading an earlier draft of this chapter

References

Abbas, A. (2002). 'Cosmopolitan de-scriptions: Shanghai and Hong Kong', in Brecken-ridge, C., Pollock, S., Bhabha, H., and Chakrabarty, D. (eds), *Cosmopolitanism*. Durham, NC: Duke.

Bell, D. (2015). 'Cottage economy', in Oakley, K. and O'Connor, J. (eds), *The Routledge Companion to the Cultural Industries*. London: Routledge.

Bond, P., DeSilvey, C., and Ryan, J. (2013). *Visible Mending: Everyday Repairs in the South West*. Axminster: Uniformbooks.

Burgess, J. (2014). 'All your Chocolate Rain are belong to us? Viral video, you tube and the dynamics of participatory culture', in Papastergiadis, N. and Lynn, V. (eds), *Art in the Global Present*. Sydney: UTSe Press

Carr, C. and Gibson, C. (2016). 'Geographies of making: Rethinking materials and skills for volatile futures', *Progress in Human Geography*, 40(3).

Cho, R., Liu, J., and Ho, M. (2018). 'What are the concerns? looking back on 15 years of research in cultural and creative industries', *International Journal of Cultural Policy*, 24(1).

Delyser, D. and Greenstein, P. (2018) 'Relighting the Castle Argyle: Remaking, restoration and the biography of an immobile thing', in L. Price and H. Hawkins (eds.) *Geographies of Making/Making Geographies: Embodiment, Matter and Practice*. London: Routledge.

Diouf, M. and Rendall, S. (2000). 'The Senegalese Murid trade diaspora and the making of a vernacular cosmopolitanism', *Public Culture*, 12(3).

Edensor, T. (2018). 'Moonraking: Making things, place and event', in Price, L. and Hawkins, H. (eds), *Geographies of Making/Making Geographies: Embodiment, Matter and Practice*. London: Routledge.

Edensor, T., Leslie, D., Millington, S., and Rantisi, N. (eds) (2009). *Spaces of Vernacular Creativity: Rethinking the Cultural Economy*. London: Routledge.

Edensor, T. and Millington, S. (2009) Christmas light displays and the creative production of economies of generosity', in Edensor, T., Leslie, D., Millington, S. and Rantisi, N. (eds.). *Spaces of Vernacular Creativity*. London: Routledge.

Edensor, T. and Millington, S. (2013). 'Blackpool illuminations: Revaluing local cultural production, situated creativity and working-class values', *International Journal of Cultural Policy*, 19(2).

Escobar, A. (2010). 'Latin America at a crossroads', *Cultural Studies*, 24(1).

Florida, R. (2002). *The Rise of the Creative Class: And How it's Transforming Work, Leisure, Community and Everyday Life*. New York: Perseus Books Group.

Florida, R. (2017). *The New Urban Crisis: Gentrification, Housing Bubbles, Growing Inequality, and What We Can Do About It*. London: Oneworld Publications.

Gibson, C, Brennan-Horley, C. and Walmsley, J. (2009) 'Mapping vernacular creativity: The extent and diversity of rural festivals in Australia', in Edensor, T., Leslie, D., Millington, S. and Rantisi, N. (eds.). *Spaces of Vernacular Creativity*. London: Routledge.

Gibson, C. and Warren, A. (2014). 'Making surfboards: Emergence of a trans-Pacific cultural industry', *The Journal of Pacific History*, 49(1).

Gilmore, A. (2013). 'Cold spots, crap towns and cultural deserts: The role of place and geography in cultural participation and creative placemaking', *Cultural Trends*, 22(2).

Gregson, N., Crang, M., Ahamed, F., Akhter, N., and Ferdous, R. (2010). 'Following things of rubbish value: End-of-life ships, "chock-chocky" furniture and the Bangladeshi middle class consumer', *Geoforum*, 41(6).

Hackney, F. (2013). 'Quiet activism and the new amateur: The power of home and hobby crafts', *Design and Culture: The Journal of the Design Studies Forum*, 5(2).

Hallam, E. and Ingold, T. (2007). 'Creativity and cultural improvisation: An introduction', in Hallam, E. and Ingold, T. (eds), *Creativity and Cultural Improvisation*. London: Routledge.

Hargreaves, I. and Hartley, J. (eds) (2016). *The Creative Citizen Unbound: How Social Media and DIY Culture Contribute to Democracy, Communities and the Creative Economy*. Bristol: Policy Press.

Hawkins, H. and Price, L. (eds) (2018). *Geographies of Making, Craft and Creativity*. London: Routledge.

Hecker, T. (2010). 'The slum pastoral: Helicopter visuality and Koolhaas's Lagos', *Space and Culture*, 13(3).

Holmes, H. (2015). 'Transient craft: Reclaiming the contemporary craft worker', *Work, Employment and Society*, 29(3).

Jackson, I. (2002). 'Politicised territory: Nek Chand's Rock Garden in Chandigarh', *Global Built Environment Review*, 2(2).

Jakob, D. and Thomas, N. (2017). 'Firing up craft capital: The renaissance of craft and craft policy in the United Kingdom', *International Journal of Cultural Policy*, 23(4).

Landry, C. (2000). *The Creative City. A Toolkit for Urban Innovators.* London: Earthscan.

Lee, J. (2010). 'Home life: Cultivating a domestic aesthetic', *Contemporary Aesthetics*, 8. http://hdl.handle.net/2027/spo.7523862.0008.015

Literat, I. and Glăveanu, V.P. (2016). 'Same but different? Distributed creativity in the internet age', *Creativity: Theories–Research–Applications*, 3(2).

Lobo, M. (2017). 'Re-framing the creative city: Fragile friendships and affective art spaces in Darwin, Australia', *Urban Studies*. DOI:10.1177/0042098016686510

Lydon, M. and Garcia, A. (2015). *Tactical Urbanism: Short-Term Action for Long-Term.* Washington: Change Island Press.

McFarlane, C. (2011). 'The city as assemblage: Dwelling and urban space', *Environment and Planning D: Society and Space*, 29(4).

McLean, H. (2017). 'Hos in the garden: Staging and resisting neoliberal creativity', *Environment and Planning D: Society and Space*, 35(1).

Mehrotra, R. (2008). 'Negotiating the static and kinetic cities: The emergent urbanism of Mumbai', in Huyssen, A. (ed), *Other Cities, Other Worlds: Urban Imaginaries in a Globalizing Age.* Durham, NC: Duke.

Miles, A. and Ebrey, J. (2017). 'The village in the city: Participation and cultural value on the urban periphery', *Cultural Trends*, 26(1).

Mommaas, H. (2004). 'Cultural clusters and the post-industrial city: Towards the remapping of urban cultural policy', *Urban Studies*, 41(3).

Mould, O. (2014). 'Tactical urbanism: The new vernacular of the creative city', *Geography Compass*, 8(8)

Ocejo, R. (2012). 'At your service: The meanings and practices of contemporary bartenders', *European Journal of Cultural Studies*, 15(5).

Platt, L. (2017). 'Crafting place: Women's everyday creativity in placemaking processes', *European Journal of Cultural Studies*. DOI:10.1177/1367549417722090.

Potts, T. (2009) 'Creative destruction and critical creativity: recent episodes in the social life of gnomes', in Edensor, T., Leslie, D., Millington, S. and Rantisi, N. (eds.). *Spaces of Vernacular Creativity.* London: Routledge.

Price, L. (2015). 'Knitting and the city', *Geography Compass*, 9(2): 81–95.

Rautio, P. (2009). 'On hanging laundry: The place of beauty in managing everyday life', *Contemporary Aesthetics*, 7. http://hdl.handle.net/2027/spo.7523862.0007.007

Robinson, J. (2006). *The Ordinary City: Between Modernity and Development.* London: Routledge.

Vannini, P. and Taggart, J. (2014). 'Do-it-yourself or do-it-with? The regenerative life skills of off-grid home builders', *Cultural Geographies*, 21(2).

Vásquez, C. and Creel, S. (2017). 'Conviviality through creativity: Appealing to the reblog in Tumblr Chat posts', *Discourse, Context & Media*, 20.

Vich, V. (2004). 'Popular capitalism and subalternity: Street comedians in Lima', *Social Text*, 22(4).

Waitt, G. and Gibson, C. (2013). 'The spiral gallery: Non-market creativity and belonging in an Australian country town', *Journal of Rural Studies*, 30.

Warren, A. and Gibson, C. (2011). 'Blue-collar creativity: Reframing custom-car culture in the imperilled industrial city', *Environment and Planning A*, 43(11).

Wendler, J. (2016). 'Grassroots experimentation: Alternative learning and innovation in the Prinzessinnengarten, Berlin', in Evans, J., Karvonen, A., and Raven, R. (eds), *The Experimental City.* London: Routledge.

Section 2

Dialogical Ecologies

3 Turning local interests into local action

Community-based art and the case of *Wrecked! On the Intertidal Zone*

Dominic Walker

Abstract

This chapter speaks to creative placemaking's heritage (Markusen and Gadwa, 2010) through the collaborative citizen-led social practice project *Wrecked! On the Intertidal Zone* (2014–16), an interdisciplinary project involving citizens, artists, and activists, drawn from Southend-on-Sea, Essex, UK, to respond to water pollution and high shipping volume in the Thames Estuary, UK.

Wrecked! used situated knowledge and citizen science drawn from local stories, histories, and locally-led initiatives to problematise the economically focused policies prioritising London, using art to promote a local agenda in local government. In doing so, it showcased how social practice placemaking (Courage, 2017a, 2017b) can be used alongside arts-led resilient practices (McKeown, 2015) to galvanise local residents, turning local interests into local action. Using Bourdieu's (1986) understandings of different forms of capital this chapter highlights how *Wrecked!*, as a community-based and artist-led initiative, recognises the value of arts beyond instrumentalising them to boost economic development as posited in creative placemaking thinking. *Wrecked!* instead utilises artistic practitioners' cultural capital, outlining the need for new forms of knowledge to redress the damage of economic capital's repeated prioritisation. By utilising different practitioners' different forms of capital, this chapter argues creative placemaking can benefit communities once local voices are heard.

Situating art in Southend-on-Sea

This chapter draws on the 2014–16 community-based project *Wrecked! On the Intertidal Zone* (2014–16) (hereafter referred to as *Wrecked!*) to explore how art can galvanise local communities into uniting around shared concerns. For *Wrecked!*, these concerns related to foregrounding Southend residents' wishes in local government as part of a bottom-up process, rather than an imposition on them of top-down national-scale agendas. These local wishes spanned cultural, social, and ecological concerns about the management of Southend, which locals felt had been detrimental. In this way, they relate to what McKeown (2015, p.24) discusses as art's role in building 'resilient places'. For McKeown, this refers to a total system of creative placemaking involving social, environmental, and commons-based placemaking embedded within an arts-led praxis for creative placemaking. This chapter argues that social practice placemaking (Courage, 2017a, 2017b) is critical to these arts-led resilient practices (McKeown, 2015) and to improving Southend-on-Sea's shared spaces, which *Wrecked!* highlights.

The origins of *Wrecked!* stem from Matsuko Yokokoji and Graham Harwood, artist-activists based near Leigh-on-Sea, a district of Southend-on-Sea with its own town council. Yokokoji and Harwood collectively form the group YoHa,[1] which form one part of a group of different people and collectives which produced *Wrecked!* collaboratively. Following a call by YoHa in the local community and local initiatives, such as the *Idea 13* initiative, and to Arts Catalyst,[2] Arts Catalyst subsequently used its social and professional connections to help commission and manage the project, sourcing funding and boosting contacts. Internationally renowned activist and tactical media collective Critical Art Ensemble (CAE),[3] known to YoHa previously, entered into discussions alongside locally engaged artists Andy Freeman[4] and Fran Gallardo,[5] local researcher Warren Harper,[6] and local curator James Ravinet.[7] Together, they were motivated to prevent two key things. First was the Thames Estuary's expansion beyond its already-excessive shipping load destined for London to service large multi-national corporations, thereby bypassing the local communities their shipping is polluting; second was the plan by former London Mayor Boris Johnson to build an island airport in the Thames Estuary, in response to the planned but problematic Heathrow expansion proposals.

This chapter uses empirical material from fieldwork in July 2015 on *Wrecked!* which draws upon residents' knowledge to consider how *Wrecked!* challenges notions of expertise, forms of capital, and context-responsive artistic practices which derive from social and spatial context. Despite appearances in previous literature (Hawkins, 2016; Lichtenstein, 2016; Worden, 2015), *Wrecked!* is a relatively under-explored yet important collaborative arts-based project, which serves to highlight several tensions between global and local issues, which shall be expressed in key junctures throughout this chapter. The chapter is divided into three parts. Part I explores the background of *Wrecked!* and its locale, Southend-on-Sea, and outlines some of the tensions arising from different ways of framing the town's space. This provides the background of the situation in Southend-on-Sea that the artistic practices in *Wrecked!* were responding to. Part II covers the four constituent parts of the project. Part III analyses *Wrecked!* through notions of expertise, capital, the wider contribution arts-based practices and engagements can make to local areas, and how such artistic forms can offer new forms of knowledge and residents' expression. It concludes by linking the chapter back to this book's key themes around creative placemaking.

Part I: Conflicting demands and the Thames Estuary

For *Wrecked!,* the artists and creative practitioners formed a community of practice with the local population from Southend-on-Sea, notably 'local ecologists, fishermen, ex-industrialists, engineers, interest groups and the general public' (YoHa, 2015, n.p.). Their goal was to uncover and showcase the complex local and situated knowledges of the Thames Estuary in the face of its changing ecology, industry, sociality, and culture. Harwood of YoHa highlights a primary motivation for producing the project, stating:

> When Boris [Johnson] goes on about his airport [island airport proposal], he goes on like no-one lives here, and because we're not that confident ... as an area, at pushing forward what we think, we need ... artists to think about how to bring that together.
>
> (Harwood, interview for Idea 13 Southend, 2014, n.p.)

For YoHa, *Wrecked!* represented a way to bring together local and situated knowledge in an area known for its ecological diversity and sensitivity, but perhaps less known for its cultural, and local histories, knowledges, and fishing methods. It drew on a diverse range of local groups with their own memories and knowledges, to enhance broader knowledge of Southend-on-Sea and preserve key cultural, historical, ecological, and local knowledge and practices.

While being Southend's home, the estuary is also a major shipping route. Copious shipping trawls through the estuary's once-fertile waters, polluting, lacerating, and churning up the waters on its way to central London. The estuary was also the focus of plans by Johnson in the early 2010s as a potential site for an island airport to alleviate the growing pressure on London Heathrow airport. It therefore sits at the confluence of a tussle between locals on the one hand and assisting London's industrial prowess on the other. This tussle relates to forms of governance and issues identified by YoHa relating to the local council while seeking to ascertain whether art can be used to promote a local agenda in local government. *Wrecked!* therefore was a project aiming to encourage local voices to unite in vocalising their sentiments about the estuary and, in doing so, to encourage locals to defend themselves in the face of increasingly polluting, detrimental, and industrialised business practices. These practices, locals argued, benefitted London to the detriment of Southend and its social, cultural, and ecological environments. The Thames Estuary and its surrounding salt marsh is also an ecologically active and diverse area. Its ecological importance, such as its attraction to rare wetlands birds like grebes, waders, geese, and ducks (JNCC, 2001), is underscored by its Special Protection Area status and conservation objectives from Natural England (UK government, 2016). The Royal Society for the Protection of Birds (RSPB) manages the area and continues to restore an ecologically sensitive area of wetlands for wild birds, touted as potentially being the biggest wetlands in Europe (RSPB, 2017). It is therefore a fertile, vibrant, and essential space of ecological importance not just for the UK, but for Europe and the wider ecological webs these ecosystems are a part of.

These ecological concerns on the one hand, and increasing globalisation through the waterways and airways on the other, represent just two framings of the complex and dynamic environment surrounding the estuary. This surrounding environment could also be framed as infrastructure for Southend's increasing leisure and tourism industry; as a container port; as historic brownfield sites; or even for its local heritage connections, such as those surrounding Anne Boleyn's house in nearby Rochford. This multitude of perspectives has a complex array of interests and influences, which *Wrecked!* hoped to foster a critical interest within and between. In this way, *Wrecked!* brought attention to the complex and often competing influences and demands of different groups with different stakes and interests, to influence governing strategies over these fragile but rich environments.

Worden (2015) argues these competing interests can be explored by creative practitioners to create open forums where local sentiment can be expressed creatively. *Wrecked!* sought to utilise these, Worden argues, stating that such a situation formed

> a gateway to reflection on the environment and a means of bringing together different groups within [the] community, as a form of open space for creativity. Extending this [the creators of Wrecked!] aim to treat the Thames Estuary as a commons; a complex shared resource. This means bringing together information that is collected by authorities with that collected informally into further events, to inform sustainable development and breakdown the bifurcation of nature and humans, and tensions between exploitation and conservation.
>
> (Worden, 2015, p.132)

As a shared resource for a shared community, *Wrecked!* invited locals to share emotional, meaningful, or otherwise unifying stories and ways of knowing their local area. *Wrecked!* sought to bring together a range of individuals with their own interests to create forums for sharing stories and create new ways of understanding and knowing the area in new forms. But how to capture these possibilities, and how might artists manage differences or tensions that can emerge, even if there is a unifying cause?

Part II: Building a community-based arts project

To approach such complex and competing demands, *Wrecked!* had four main parts, each of which was a separate artistic commission, the details of which I discuss in this section. Each of these parts drew on different actors from the collective of locals, practitioners, and artists. These four parts sought to stimulate residents' interest in the project through different forums, which would then inspire engagement with the wider issues *Wrecked!* explored. These four artistic commissions sought to focus on each of the project's four key themes. These themes concerned the local ecology in *Talking Dirty*;[8] resident participation in sustainably looking after the Thames Estuary in *Citizen Science*;[9] historical and local cultures in *Graveyard of Lost Species*;[10] and historical and local knowledges and practices in *Line in the Sand*. I now discuss these four parts and their themes.

The first part of *Wrecked!* was *Talking Dirty: Tongue First!* (Arts Catalyst, 2017d; hereafter referred to as *Talking Dirty*), a series of public events and workshops revolving around local foodstuffs. Led by local artists Gallardo and Freeman, and environmental chemist and food scientist Mark Scrimshaw, *Talking Dirty* used foodstuffs grown in and around the Thames Estuary to prepare tasty food dishes. *Talking Dirty* called on locals to share recipes based on local produce, foraging practices, and tips, as well as stories relating to local food. From these, dishes would be prepared for visitor sampling at each event. These recipes were compiled into a recipe book which residents could cook from at their leisure. The array of local species meant some could also be smoked, the inhalation adding a form of embodied engagement with the artwork. For Worden (2015), these practices employed as part of *Wrecked!* helped make the environment 'visible' through taste or smell, such as from the local flora or through ingredients in vaporisers. The embodied practice of consuming the estuary's produce represented a visceral form of knowing, stoking interest and providing a different way of understanding the environment of this flora. However, these workshops also raised key ecological difficulties. Given the Thames Estuary's importance for the UK's shipping industry, this contribution has come at a price. Many of the foodstuffs garnered from the estuary are not fit for human consumption and decades of daily mass pollution has produced toxic levels of some heavy metals, minerals, and other constituents in these organisms. This provided a stark reminder of the ecological damage already caused by industry and its implications further down the food chain.

To further highlight Southend's environmental damage and encourage action, Freeman and Scrimshaw led citizen science workshops in conjunction with *Talking Dirty*. These workshops were titled *Citizen Science* (Arts Catalyst, 2017e) and traced the practices of waste disposal at the nearby Two Tree Island in Leigh. This was the second part of *Wrecked!*. Two Tree Island was formerly a sewage works and then a landfill site prior to its current running by the Essex Wildlife Trust. It was therefore an ideal location for *Citizen Science*, drawing local residents together to develop their understandings of the estuary in ways understandable to them, and that could be monitored by them. One focus of *Citizen Science* was on how industry and scientists use sensors in the estuary to chart and monitor particular indicators such as tidal flows, levels of certain pollutants, humidity, and wind speed. These data are relayed in conjunction with the growth of particular aquatic species, such as cockles, types of fish, and barnacles, to form an overview of how the factors relate to the species. Crucially, both the sensors and the species levels are used by scientists when discussing with politicians, policymakers, or legal professionals the implications of controversial practices such as dredging. *Citizen Science* introduced these ideas to local individuals to highlight science's reduction of a complex and sensitive ecological area to metrics with limited applicability. For the collaborative practitioners producing *Wrecked!*, a reduction of the area represents a misunderstanding and mis-handling of this delicate ecosystem. These *Citizen Science* workshops therefore sought to use local understandings of the estuary to question these methods, offering ways for the public to form and subsequently disseminate understanding of their environment in ways they can relate to, monitor, and engage with. *Citizen Science*, then, sought to inspire local interest not just in the *Wrecked!* project, but in the ecological area's future, encouraging a renewed sense of stewardship for the local environment on locals' terms and understandings.

The third part of *Wrecked!* was the *Graveyard of Lost Species* (Arts Catalyst, 2017c), which centred on a restored ship rescued from the estuary mud. The ship, named the *Souvenir*, was a 40-foot Thames fishing boat with on-board shrimp boiler (otherwise known as a 'bawley'), weighing 12 tons. *Souvenir* was cleaned by YoHa and restored in a public setting to encourage questions and engagement from locals. The *Graveyard of Lost Species*, like the previous two parts, also sought situated knowledges from locals. It attempted to construct a monument to previous wildlife species, marine creatures, landmarks, livelihoods, fishing methods, and dialects that historically thrived in the estuary but have since declined or become extinct. The artists led enquiries with Southend residents to garner local expertise relating to these 'lost species' (Arts Catalyst, 2017c). The *Wrecked!* practitioners encouraged locals to contribute their stories, knowledge, and memories of the local area, which would then be emblematised on *Souvenir* as a monument by Southend's people, for Southend's people. Key words, chosen by locals, would be laser-cut into the wood of the ship, reminding locals of what Southend used to be and have. In this way, *Souvenir* represented how Southend had changed from a distinctive town with its own cultural phenomena to one where its identity was becoming diluted through the impact of London's increasing globalisation. Southend residents were therefore concerned for Southend's welfare in several ways. As pollution from enormous tankers and barges in the estuary grows, so species decline, become vulnerable in changing food webs, and become extinct. Landmarks are removed to make way for tourism or business developments and livelihoods of local fishermen and women are jeopardised by decimated stocks or cheaper labour elsewhere. These changes have gradually progressed through time – often unnoticed – and the *Souvenir* represented one way to awaken locals' sensibilities to these by displaying them in Southend's marshes. Here, they formed part of the local landscape once more. Accordingly, the *Souvenir* also gestured towards Southend's future as well as its past, implicitly questioning how the town might, in time, further change with new stories, memories, food webs, and dialects created.

The final part of *Wrecked!* was *Line in the sand* (Arts Catalyst, 2017b),[11] an ephemeral public artwork which used writing in the Thames Estuary sand to communicate with London-bound planes. The shifting mud of the estuary underlies the flight path to London's airports, highlighting the impact of the continuous stream of commercial and passenger jets overhead. To produce *Line in the sand*, YoHa collaborated with local Southend fishermen and individuals to choreograph boats whose arrangement spelt out messages. These messages were indented by trawlers, sifting sand two feet wide and two inches deep, over which water lays and shimmered in the sunlight. The content of the messages was intended to draw from research undertaken as part of the *Graveyard of Lost Species* and was decided upon following consultation with the public (Arts Catalyst, 2017b). A key factor in determining *Line in the sand* was to be attentive to and showcase the local histories, knowledge, and culture of Southend's declining fishing industry and the fishermen and women whose livelihoods depend on them. These livelihoods, residents argued, were being increasingly usurped by air cargo, which now transports over five per cent of the planet's annual sea catch (Arts Catalyst, 2017b), flying it in from cheaper sources around the world. Writing in the sand using fishing boats, then, represented a fitting way to honour the declining culture historically seen as a lynchpin of Southend's cultures and livelihoods.

When combined, these four component parts of *Wrecked!* therefore represented an opportunity to stimulate interest in these areas of Southend's culture, with a view to opening up debates on key tensions in several ways, which the following section will address.

Part III: Local engagement for local change

Wrecked!, then, highlighted four key facets of context responsive artistic practice. First, it responded to the top-down imposition of neoliberalism-driven national policies prioritising business and commerce above all. To argue this point, I now engage with Bourdieu's (1986) recognition of the different forms of capital, highlighting how economic capital has previously dominated over other forms of capital in ways *Wrecked!* draws attention to. *Wrecked!* highlighted the clashing understandings about which forms of capital should be prioritised, played out alongside the local versus national scale. In Southend, the national-level policies have favoured, in Bourdieu's (ibid.) terms, economic capital at the expense of cultural and social capital. In other words, they favoured capital exchanges which convert into economic resources, such as money, above all else. Over decades, this pursuit has led locals to reject this economic focus after witnessing its detriment to their area through ecological damage, dialect loss, and the decline of local fishing methods, local stories, and histories in the name of profit, a profit that Southend sees little, if any, of. The daily trail of shipping containers trawling down the Thames Estuary packed with London-bound goods, polluting and decimating Southend's delicate ecology, if left unchecked, further disrupts the complex and delicate ecosystems the Southend people and wildlife rely on, such as the wetlands birds the RSPB are attempting to safeguard. Other forms of damage have implications for local food sources, now rendered inedible by the toxicity levels in the estuary.

Residents in Southend showed in *Wrecked!* that they instead favoured a form of arts-led resilient practices (McKeown, 2015), moving away from the previous economic focus and towards what Bourdieu (1986) terms cultural and social capital, i.e. forms of capital which underpin notions of value. Cultural capital involves practices, materials, and values which help produce a shared identity among those sharing similar positions. Cultural capital therefore expresses 'value' through materiality, corporeality, or institutions. Social capital, meanwhile, relates to human relationships among social groups. It focuses on reciprocity within the arts/social and situated practice, such as that highlighted by McGonagle (2010, 2011), and uses interactions in these groups as currency to determine their relative capital. 'These relationships', Bourdieu (1986, p.248) writes, 'may exist only in the practical state, in material and/or symbolic exchanges which help to maintain them'. Social capital, then, expresses 'value' through human relationships and social interaction. For these reasons, Southend residents used *Wrecked!* to show a need for cultural and social capital over economic.

My second point relates to how *Wrecked!* engaged with cultural and social capital. In *Wrecked!* creative practitioners and artists attended to unlocking Southend's collective cultural capital, but not through the creative placemaking approach (Markusen and Gadwa, 2010) often seen in urban areas. This is possibly for two reasons. First, Southend's residents insist it does not need rejuvenating, or at least the kind of regeneration offered by economic-led initiatives. Second, creative placemaking must be approached carefully. When portrayed as creative practitioners being parachuted into unfamiliar urban centres to stimulate economic development, creative placemaking could be seen as prioritising economic capital. Bourdieu notes how cultural capital can be 'convertible, on certain conditions, into economic capital' (1986, p.242), and it could be argued that practitioners' cultural capital is sought out for its conversion into its economic potential rather than to culturally improve the centres and their quality of life. Consequently, creative placemaking must be utilised for the right purposes in the right situation to avoid undermining the non-economic value of creative practitioners.

Wrecked! acknowledged these difficulties, utilising creative practitioners' cultural and social capital rather than economic potential. Accordingly, *Wrecked!* encouraged residents to engage with artists and creative practitioners to help express their opinions, in a town historically too accepting of detrimental practices, according to YoHa's Harwood. 'Because we're not that confident ... as an area, at pushing forward what we think', Harwood asserted as reported earlier in this chapter, 'we need ... artists to think about how to bring that together' (Harwood, interview for Idea 13 Southend [2014, n.p.]). For Harwood, artists therefore provide an opportunity to convert interest into action, using 'their methodology to engage people differently' (ibid.). They can encourage expression of locals' opinions in appropriate ways, namely the fostering and bringing together of different forms of expression into a shared opinion. This shared opinion adds weight when considering decisions in local government; a first step on a long journey to overturning decades of prioritising the economic. Artists, therefore, can facilitate residents' self-empowerment to cultivate a sense of shared identity, and push an agenda into local-level politics. In Bourdieu's terms, artists and creative practitioners in *Wrecked!* used their cultural capital to improve the standard of living, shared spaces and shared identity by treating residents as equals and pushing the local council to reflect this by focusing on more than just economics. Involving creative practitioners in the placemaking process, then, highlights two points. Firstly, the difficulty of ensuring that creative practitioners in the placemaking process are used for their cultural capital, rather than as vehicles to boost economic capital. Secondly, how social practice placemaking (Courage, 2017a, 2017b) can be an effective way to ensure the use of creative practitioners' cultural capital. This relates to the third point I now attend to.

Social practice placemaking (Courage, 2017a, 2017b) draws on social, participatory, and interdisciplinary arts practices to integrate specific knowledge cultures which offers residents and creative practitioners equal expert status. Using social arts practice gives power to visitors or audiences of artworks, allowing them to contribute to shaping or informing the artwork (Jordan, 2017; Walker, 2017). In social practice placemaking, visitor power extends beyond the artwork. Visitors from the community, such as in *Wrecked!*, became participants in the development of the artwork. Artists and the community were recognised as 'equal experts' of local and situated knowledge to help improve the community area (Courage, 2017b). One example of this community improvement from *Wrecked!* lay in plans announced, during the project, for a third runway at Heathrow airport rather than pursuing Boris Johnson's island airport idea (Nizinskyj, 2015). This represented a substantial victory for Southend, who had strongly opposed the island airport proposals. Social practice placemaking therefore appreciates the contribution of those possessing cultural capital, such as creative practitioners, and those of social capital, such as residents. Accordingly, *Wrecked!* used a blend of local, experienced, and externally sourced creative practitioners to vocalise local and situated knowledges and local opinion, which clearly lambasted top-down decisions against locals' interests.

As *Wrecked!* showed, participatory projects can galvanise local communities into action, drawing on their shared identity of their town and its public spaces to instead argue for bottom-up approaches. Using different local cultural facets helped to underscore the importance of giving residents equal authority in decision-making for their local spaces. In this way, social practice placemaking becomes an effective means of bringing together artists and residents to improve the local area's public spaces. In the case of *Wrecked!*, artists used their cultural capital to stimulate interest among locals who bore social capital. The artists then translated this interest from locals into action, seeing themselves and locals treated as equals among 'experts' in decision-making processes. Artists can therefore help highlight the value of different forms of capital beyond the economic.

Using creative practitioners to engage with residents to elicit local knowledges also highlights a fourth and final theme of *Wrecked!*, which is that of citizen science. As this chapter has explored, artists can help produce, encounter, and validate forms of cultural and social capital, garnering expressions of them through different means. In *Wrecked!*, this meant recognising different forms of knowledge emerging from the project's four main components, locking into other 'forms of citizen science that maybe produces other kinds of knowledge'

(Harwood, interview for Idea 13 Southend [2014: n.p.]). For Harwood, *Wrecked!* represented an opportunity to utilise citizen and situated knowledges in new and creative ways to produce different ways of re-imagining knowledge which could improve Southend as a shared, public space. These forms of knowledge drew on existing and local sources rather than established, certified 'experts'. Hawkins (2016) argues that projects such as *Wrecked!* combine 'participatory practices with citizen science'. Accordingly, Hawkins argues that the result of *Wrecked!*, and of many other projects is to recognise creative practice as something that can contextualise the scientific expert. In doing so, creative practice allows other voices to be heard while recognising the value of other practices in how scientific knowledge is made. For Hawkins, participatory art, then, can work to expand the notion of 'expert' and draw on locals' situated knowledge to consider them as equals. This approach widens notions of expertise, acknowledging the cultural and social capital residents possess, and making it comparable to educational qualifications in treating them as 'experts'.

However, with *Wrecked!*, residents are not just contextualising scientific experts, but also the politicians, economists, and other professionals who repeatedly favoured economic capital at the expense of everything else. In particular, *Wrecked!* focused on local councillors who, residents felt, had consistently put their residents' needs behind London's. Alongside these were the politicians and economists who suppressed attempts at replacing economic capital with other forms of capital and allowed economic capital to dictate policy and governmental decisions. These decisions impacted at the local level and went against Southend's residents' wishes. Citizen science formed one part of residents' riposte to these top-down impositions, with residents deciding what is important to them, rather than being told what to value. These top-down approaches fundamentally mis-understood how Southend residents related to their environment, removing mani-festations of local culture for the sake of finance. Accordingly, as an approach, citizen science re-conceives how local forms of knowledge are encountered. Local knowledge becomes cherished rather than overlooked in favour of certified 'experts' parachuted into an unfamiliar area with little knowledge of its unique intricacies. In *Wrecked!* artists helped re-imagine and re-present existing forms of knowledge in new ways drawn out from local histories, and then helped install these as bases of accredited and accepted forms of knowledge. *Wrecked!* formed an output, which can be shown to local government as a demonstrable example of locals' concerns, a 'transformative politics of the local' (Worden, 2015, p.132).

In this way, *Wrecked!* highlighted how context-responsive artistic practices can produce new forms of knowledge and social practice placemaking. This pre-empts Zebracki and Palmer's (2018) call for questions, answers, and engagements (Kwon, 2004) around artworks dependent on their spatial context. I have pre-viously attended to some of these questions, answers, and engagements in a short commentary demonstrating how artworks based around such spatial context can re-present previously overlooked forms of knowledge (Walker, 2015). In both Zebracki and Palmer's (2018) and my (2015) works, the spatial context extends into the social, an idea which *Wrecked!* also drew on. Spatially, residents' particular rhythms, practices, and knowledges are engaged with, produced, and re-enacted. These spaces are socially practiced (Zebracki and Palmer, 2018) through social engagements and revolve around the spaces (Lefebvre, 1996) and asso-ciated practices of everyday life (de Certeau, 1984). These spaces and practices of everyday life are then played out across different scales, implicating Southend in national and international networks. These spaces and practices necessarily entail different aspects of the community and its situated knowledge, which the four parts of *Wrecked!* highlighted. These practices related to the shared community understanding of Southend as a whole, and of places-in-the-making (Silberberg et al., 2013), continually re-made through their everyday practices. Southend's residents decide how they use their own spaces by how they *act* in them. By acting in the spatial, such as by attending and contributing to *Wrecked!*, citizens acted to give the project traction, solidifying community bottom-up shared under-standings and meanings of these local areas.

Conclusion

Wrecked! is an example of artists drawing on local histories, cultures, and knowledges as part of context-responsive projects to improve local areas ecolo-gically, culturally, and socially. Accordingly, the project speaks to the heritage of creative placemaking, highlighting the need for equal status of locals amid the complexities and tensions of each individual (geographical) area, each individual project, and each unique form of local knowledge. This shows the importance of social practice placemaking in finding new ways to improve the quality of life in urban centres. Using social practice placemaking allows locals to be sought out for key ideas and knowledge rather than them having to respond to imposed policies and forms of knowledge. In this way, *Wrecked!* highlights the effective-ness of combining social practice placemaking with arts-led resilient practices (McKeown, 2015), giving residents control of managing and improving their living spaces. Fundamentally, *Wrecked!* shows the importance of valuing local contributions and for ascertaining local opinion through different forms of expression to re-imagine and re-present different forms of knowledge. Together, these are elicited by artists and creative practitioners motivated for change, and whose contribution helps residents express their sentiments and cultural heritage. In this way, *Wrecked!* recognises the value of arts beyond instrumentalising them to boost economic development, instead utilising their cultural capital and out-lining the need for new forms of knowledge to redress the damage of decades of elevating economic capital above all other forms.

Notes

1 YoHa, http://yoha.co.uk/about
2 Arts Catalyst, www.artscatalyst.org
3 Critical Art Ensemble, http://critical-art.net
4 Andy Freeman, www.artscatalyst.org/artist/andy-freeman

5 Fran Gallardo, www.artscatalyst.org/artist/fran-gallardo
6 Warren Harper, www.artscatalyst.org/artist/warren-harper
7 James Ravinet, www.rca.ac.uk/students/james-ravinet
8 *Talking Dirty: Tongue First!*, www.tonguefirst.com
9 *Citizen Science*, http://wrecked.artscatalyst.org/landing/4
10 *Graveyard of Lost Species*, http://wrecked.artscatalyst.org/landing/2
11 *Line in the sand*, http://wrecked.artscatalyst.org/landing/3

References

Arts Catalyst. (2017a). *About Arts Catalyst* [online]. Available from: www.artscatalyst.org/content/about-arts-catalyst [Accessed: 5 July 2017].

Arts Catalyst. (2017b). *Line in the sand* [online]. Available from: http://wrecked.artscatalyst.org/landing/3 [Accessed: 11 August 2017].

Arts Catalyst. (2017c). *Graveyard of Lost Species* [online]. Available from http://wrecked.artscatalyst.org/landing/2 [Accessed: 12 August 2017].

Arts Catalyst. (2017d). *Talking Dirty: Tongue First!* [online]. Available from www.tonguefirst.com/home [Accessed: 20 October 2017].

Arts Catalyst. (2017e). *Citizen Science in the Thames Estuary.* [online]. Available from http://wrecked.artscatalyst.org/landing/4 [Accessed: 20 October 2017].

Bourdieu, P. (1986). 'The forms of capital', in Richardson, J. (ed.). *Handbook of Theory and Research for the Sociology of Education*. New York: Greenwood.

Courage, C. (2013). 'The global phenomenon of tactical urbanism as an indicator of new forms of citizenship', *Engage in the Visual Arts*, 32.

Courage, C. (2017a). 'Art practice, process, and new urbanism in Dublin: Art Tunnel Smithfield and social practice placemaking in the Irish capital', *Irish Journal of Arts Management and Cultural Policy*, 4.

Courage, C. (2017b). *Arts in Place: The Arts, the Urban and Social Practice*. Abingdon: Routledge.

De Certeau, M. (1984). *The Practice of Everyday Life*. Trans. Rendall, S., London: University of California Press.

Hawkins, H. (2013). 'Geography and art: An expanding field – site, the body and practice', *Progress in Human Geography*, 37(1).

Hawkins, H. (2016). *Creativity*. London: Routledge.

Idea 13 Southend. (2017). *Idea13 TV: Wrecked on the Intertidal Zone, Graham Harwood (Yoha), Fran Gallardo, and Claudia Lustra* [online]. Available from www.youtube.com/watch?v=T5DJPqsImvs [Accessed: 5 July 2017].

Joint Nature Conservation Committee. (2001). *SPA Description: Thames Estuary and Marshes* [online]. Available from http://jncc.defra.gov.uk/page-2042 [Accessed: 17 July 2017].

Jordan, C. (2017). 'Joseph Beuys and Social Sculpture in the United States'. Ph.D. thesis, City University of New York.

Kwon, M. (2004). *One Place after Another: Site-Specific Art and Locational Identity*. Cambridge, MA: MIT Press.

Lefebvre, H. (1996). *Writings on Cities*. Trans. and ed. by Kofman, E. and Lebas, E. Oxford: Blackwell.

Lichtenstein, R. (2016). *Estuary: Out from London to the Sea*. London: Hamish Hamilton.

Markusen, A. and Gadwa, A. (2010). *Creative Placemaking*. Washington, DC: Mayors' Institute on City Design and the National Endowment for the Arts [October 2010].

Available from www.terrain.org/columns/29/CreativePlacemaking_NEA.pdf [Accessed: 7 July 2017].

McGonagle, D. (2010). *Passive to active citizenship: A role for the arts*. Bologna in context conference, 24 October 2010, The Honourable Society of King's Inn, Dublin.

McGonagle, D. (2011). 'An "other" proposition – situating reciprocal practice', in Parry, B., Tahir, M., and Medlyn, S. (eds.), *Cultural Hijack: Rethinking Intervention*. Liverpool: Liverpool University Press.

McKeown, A. (2015). 'Deeper, slower, richer: A slow intervention towards resilient places', *Edge Condition*, 5.

Nizinskyj, P. (2015). 'New Heathrow runway "good for Southend" in *Basildon, Canvey & Southend Echo*, [online], 3rd July 2015. Available from: www.echonews.co.uk/news/13367946.New_Heathrow_runway__good_for_Southend_/ [Accessed 23 October 2017].

Royal Society for the Protection of Birds. (2017). *Wallasea Island Wild Coast Project* [online]. Available from https://ww2.rspb.org.uk/our-work/our-positions-and-casework/casework/cases/wallasea-island [Accessed: 17 July 2017].

Silberberg, S., Lorah, K., Disbrow, R, Muessig, A., and Naparstek, A. (2013). *Places in the Making: How Placemaking Builds Places and Communities*. White Paper. Cambridge, MA: MIT, Department of Urban Studies and Planning. Available from http://dusp.mit.edu/sites/dusp.mit.edu/files/attachments/project/mit-dusp-places-in-themaking.pdf [Accessed: 17 July 2017].

UK government. (2016). *Thames Estuary and Marshes Special Protection Area: Site Information (Draft)* [online]. Available from www.gov.uk/government/publications/marine-conservation-advicefor-special-protection-area-thames-estuary-and-marshes-uk9012021/thames-estuary-and-marshesspa-site-information-draft [Accessed: 17 July 2017].

Walker, D. D. (2015). 'Atomic Age Rodents: In search of the first animals of the Anthropocene', *Society and Space* [online], 20th August 2015. Available from: http://societyandspace.org/2015/08/20/atomic-age-rodents-dominic-walker/ [Accessed 17 July 2017].

Walker, D. D. (2017). 'Experimental Geographies, Artists, and Institutions: Spaces of and Practices for Knowing'. Ph.D. thesis, University of Exeter.

Worden, S. (2015). 'The Earth sciences and creative practice: Entering the Anthropocene', in Harrison, D. (ed.). *Handbook of Research on Digital Media and Creative Technologies*. Hershey, PA: Information Science Reference.

YoHa. (2015). *Wrecked on the Intertidal Zone* [online]. Available from http://yoha.co.uk/wrecked_ [Accessed: 5 July 2017].

Zebracki, M. and Palmer, J. (eds.). (2018). *Public Art Encounters: Art, Space and Identity*. Abingdon: Routledge.

4 Arrivals and departures

Navigating an emotional landscape of belonging and displacement at Barangaroo in Sydney, Australia

Sarah Barns

Abstract

This chapter offers a practitioner perspective on creative placemaking within a contested landscape of urban renewal. Focused primarily on a public art project developed by Esem Projects in 2015 at Barangaroo, Sydney, the chapter discusses the tensions involved in negotiating contested territories of historical, institutional and community attachment to a prime waterfront precinct. Through creative practice, the resources of memory and affective engagement were used to expand the different layers of meaning ascribed to the place, many of these now erased from the physical landscape through the process of urban renewal. In this context, conjuring an emotional landscape of attachment became an act of resistance to urban revitalisation, while at the same time renewing, celebrating, and expanding the many versions of place that have existed in this significant waterfront precinct through time.

Introduction

Over the past decade we have seen the value of place become increasingly central to the arts of urban renewal. It is now widely recognised that when a place has a strong identity, a strong 'sense of place', this brings with it not only improved social cohesion and community benefits, but also uplifts in land value. Placemaking now defines a thriving ecosystem of practitioners, consultants and designers working in their respective fields of architecture, urban design, property development, local government, community engagement and public art. Each is united in working towards the common goal of making great places. By employing arts-led methodologies to support community programming, events, festivals and installations, creative placemaking plays a critical role in helping to build and enhance urban vitality and social connectivity at the local level.

In Australia, the funding environment for creative placemaking has been less reliant on philanthropic investments than has been the case in places such as the United States (Wilbur, 2015). Major local government organisations such as the City of Sydney and the City of Melbourne invest significantly in creative placemaking practice through a mix of public art programs, local events and festivals. Support from New South Wales (NSW) agencies such as Destination NSW[1] for major events such as Vivid Sydney,[2] attracting over two million visitors each year, has created a platform for a range of media artists and designers to exhibit and showcase works using the built environment of Circular Quay as their canvas. Many creative placemaking initiatives have also emerged as a result of significant urban transformation projects underway across metropolitan areas. In Sydney, this includes major transport infrastructure investments and ambitious urban renewal projects, which are each accompanied by new creative placemaking investments designed to help 'activate' new precincts. Across the built environment sector there is also growing awareness of the importance of programmatic responses to place activation, which incorporate community-focused temporary activations just as often as 'one-off' permanent public art commissions.

In this environment, the avenues of support for creative work and programming in the public domain have multiplied. By turning to the public realm as a site of creative practice, creative arts practitioners can reduce their reliance on more traditional – and diminishing – sources of arts funding, while exploring diverse and emerging art practices that incorporate dialogue between people, place, community and the dynamics of urban transformation. Many diverse creative practitioners are now engaged in the field, from traditional public artists or community arts organisations to commercial designers, advertising agencies and landscape, lighting and interaction designers.

There is, of course, a clearly instrumental logic behind the funding that flows into creative placemaking practice. Creative artworks or events need to align with a largely celebratory ethos that champions and enhances the connections between people and place. Situated within the wider ethos and practices of placemaking, the work of creative placemakers is designed to support and facilitate a range of affective and emotive responses to place. In doing so, creative placemakers affirm the expression of place as a powerful expression of culture. As Place Leaders Asia Pacific has suggested: 'Engendering a sense of "place" and making of great public places takes an alchemy of elements – activity, material, structure, purpose and inspiration. Great places are memorable and stimulate emotional responses in people' (Place Leaders, 2017).

However, what many of these celebratory accounts of place and its emotional entanglements often fail to accommodate are the many ways in the relationship between place, affect, memory and public culture can often be highly fractured and discordant. Defining what constitutes a place, and what aspects of it should be celebrated, and by whom, can, at certain sites, be a highly political act. In this context, if we acknowledge that notions of place are always subjective, contested and evolving, the work of the creative placemaking practitioner will in turn necessitate negotiations between competing claims and narratives of place.

In this chapter I discuss this work of negotiating multiple contested place narratives as it shapes creative placemaking practice. The discussion, which springs from a body of work I created over several years within the Sydney waterfront precincts of Millers Point and Barangaroo, reflects on the different frameworks of investment, attachment and value associated with one of Sydney's major urban renewal projects. As I discuss, the conjuring of an emotional landscape, of memory and belonging, through a site-specific response that incorporated community archives, audio-visual recordings and poetry, became a way to resist a contested narrative of revitalisation that obscured just as much as it 'activated'. While not resolving any of the tensions that have simmered and often erupted during the course of the precinct's development, it was by creating an intimate, experiential sense of the depth of personal connections that had shaped and been shaped by this place that I sought to enable multiple forms of attachment to be acknowledged, while also allowing new connections to be forged.

Welcome to Barangaroo

In 2015 Esem Projects,[3] a Sydney-based media arts and design practice that I co-direct, was invited to produce a major installation to mark the opening of Barangaroo Reserve. The installation was to be established in what was once a container terminal in the former East Darling Harbour, is Sydney's newest public parkland which was created as part of a controversial AUD $6bn redevelopment of the waterfront lands of East Darling Harbour. Our installation, titled *Arrivals and Departures* (2015),[4] was specifically commissioned to celebrate and acknowledge the maritime history of the area as a working port (Figures 4.1 and 4.2).

The commission reflected interest in and support for our creative arts practice as one that combines elements of creative placemaking with heritage interpretation

Figure 4.1 Families using the vintage typewriter in *Arrivals and Departures*, Barangaroo Cutaway, 2015. Credit: Esem Projects.

Figure 4.2 Former residents and maritime workers projected within Barangaroo Cutaway, with shipping containers as 'Storyboxes'. *Arrivals and Departures*, Barangaroo Sydney, 2015. Credit: Esem Projects.

and digital design. Working with public collections and local community story-tellers, we create localised 'living archives' of place comprising oral histories, documentary films and photographic collections. We use site-specific installations and experience design methods to curate and reinscribe these 'trace' recordings within the built environment. Like other creative placemaking practices, our commissions often come from local councils and place managers who recognise the potential to incorporate community and historical storytelling into the built environment through temporary installations and projection media.

Our commission for Barangaroo was one of three installations featured as part of a three-month program of 'Welcome Celebrations'. Typical of many creative placemaking programs, the Welcome Celebrations featured a series of curated performances, installations, food events and public talks that together created a broad range of attractions for Sydney audiences to visit this new harbourside precinct for the first time. Each of the three installations were asked to offer a creative response to different symbolic and historic representations of the place: 'Stone' for indigenous representations, 'Sea' for the maritime history and 'Sky' as representatives of future hopes and expectations for the site.

We were honoured to be invited to produce an installation for the opening of this significant new precinct, specifically focused on the theme of 'Sea' and post-settlement maritime history. Yet, we knew there would be tensions to resolve. On the one hand, the commission gave us the opportunity to engage large audiences with the significant maritime history of Sydney's East Darling Harbour. The opportunity to interpret the complex social, industrial and environmental history of the site as a maritime precinct was incredibly exciting. The exhibition space itself was remarkable: a cavernous new cultural space called 'The Cutaway' in central Sydney, enabling us to work at a scale far larger than we had known before.

It was, however, quite evident that the revitalisation of the precinct, and the act of 'welcoming' new audiences to the transformed headland, was also, for some, an act of banishment. The redevelopment of Barangaroo has pit developer interests – in this case the interests of the NSW state government – directly against the needs of the local community. The management of the site's industrial heritage has also provoked widespread public outcry over the very role and place of history in the wider narrative of Sydney as a city. Accepting the commission, we knew, would necessitate negotiating directly with these tensions.

Before describing our creative response to these tensions, I want to first capture the dynamics of debate and conflict generated by this project of urban revitalisation – in short, the contested terrain of place that would shape the design and activation of our project.

Revitalisation as displacement

The largest urban renewal project in Sydney since the 2000 Olympics, Barangaroo covers some 22 hectares (220,000sqm) of former industrial land along the harbour and lies adjacent to the central business district. It includes a new financial district comprising three commercial office towers designed by Richard Rogers; an AUD$1.5b casino; a series of high-end residential apartments, as well as shopping, dining, hotel, and hospitality services and a public promenade leading to the new headland reserve.

The many controversies provoked by the redevelopment are familiar to most Sydney-siders. The winning urban design for the development, selected in 2005 through an international urban design competition, was ultimately set aside by planning authorities following a series of modifications to the masterplan, which included a doubling of the allowable density and significant increases in the proposed height of buildings. Influential former Prime Minister Paul Keating, who sat on the design review panel for the development, negotiated the inclusion of a naturalistic headland (the Reserve), intended to remove references to the precinct's maritime history, thereby returning the foreshore to its pre-colonial condition.

These modifications raised the ire of the architectural profession, aghast at a process that had seen a competition-winning masterplan scuttled through a series of unaccountable interventions taking place at a highly significant site alongside Sydney's iconic harbour. The outrage was not simply on account of the poor adherence to transparent decision-making: many also saw a complete lack of sensitivity to the site's significant industrial heritage (Drew, 2015; Reinmuth, 2012; Weller, 2010).

The governance of the development likewise attracted intense criticism. The project was listed as a 'State Significant Site' and subsequently exempt from many conventional planning processes. The creation of a new state government-owned agency, the Barangaroo Delivery Authority,[5] provided a governance mechanism to reduce the influence of the City of Sydney in the planning and development process. The placename has also provoked criticism: Barangaroo, the partner of Australia's first Aboriginal 'ambassador', Bennelong, opposed the theft of native land and belonged to the Gamaragal people of Manly. She was not of the Eora people who occupied the area when colonialists first arrived, and whose own placenames were first documented by English linguist marine Lieutenant William Dawes. The Aboriginal land council has refused to endorse the name.

The continuing and fractious public debates sparked at many stages of the development of Barangaroo have returned the city to familiar battles over the fate of its heritage and lack of trust in the city authorities, whose capacity to rule in favour of developer-led interests continues to shock and dismay. The contributions to the development made by noteworthy international architects, no less than Lord Richard Rogers, have heightened cynicism towards the values of urban elites. Indeed, Rogers' own words seem to underscore this cynicism:

> The city has been viewed as an arena for consumerism. Political and commercial expediency has shifted the emphasis of urban development from meeting the broad social needs of the community to meeting the circumscribed needs of individuals. The pursuit of this narrow objective has sapped the city of its vitality.
>
> (Drew, 2015, para. 15)

To Reinmuth (2012, para. 24), the experience of Barangaroo made plain that city leaders continue to persist in prioritising 'the business of consuming the city rather than making the city with the involvement of its citizens', and at a site like Barangaroo, 'we will only continue to distrust our leaders and the places they impose on it'. Drew (2015, para. 18) writes that '[l]ike a giant finger given to Sydney, everything about it is selfish and narcissistic: its excessive height, public exclusion, and monopolisation of harbour views for a wealthy few'.

Yet, despite the very public controversies sparked during the development of Barangaroo, others found much to praise as the project neared completion. Australian environmentalist and writer Tim Flannery, writing for the *New York Review of Books* (2017), celebrated the opening of the Barangaroo Reserve as an 'act of restitution'. Flannery (ibid., para. 1) has compared the opening of Sydney's Barangaroo Reserve in August 2015, what he calls 'an expansive, charming public space at the heart of a great commercial city', with New York's High Line[6] and London's East End Olympic[7] redevelopment, each landmark public parks that 'help define a major metropolis's sense of place'.

For Flannery, trained as an environmental scientist, the qualities of biodiversity and ecology now embraced at the site pay respect to the area's pre-colonial environmental history. Barangaroo has provided the opportunity to restore the late eighteenth-century shoreline of the harbour, which was altered – what Keating called an act of 'vandalism' (Legge, 2015) – during the industrialisation of the harbour. By removing the containerised port infrastructure and replacing it with public parkland that incorporates many of the plant species native to the area, the making of Barangaroo has also made open to the public a perspective on the harbour lost for a century. Referring to the parkland's extensive use of natural sandstone, Flannery (2017, para. 5) praises the development for 'restoring [Sydney] standstone to the prominence it deserves.' Indeed, to Flannery (ibid., para.14), the whole development 'marks a turning point in the relation of Sydney's people with its past.'

To those who have opposed the development, and specifically the public parkland, the re-creation of a long-lost headland represents a kitsch and 'explicitly phoney naturalism' that has sent Australian landscape architecture 'reeling back to the eighteenth century' (Weller, 2010, para.9). Critics of the natural headland believe the industrial harbour has significance with regard to Australia's historical development as a nation and aspects of its character should have been preserved. Indeed, it is from the docks and wharves that once dotted this area of the harbour that Australia's exports were first shipped; it was also the embarkation point for the millions of migrants who first arrived in Australia; it was the disembarkation point for many soldiers departing during the wars of the early twentieth century; and it was a place where one of the world's great maritime unions learned to fight for its workers' rights. Returning to a pre-settlement landscape means effectively erasing the physical markers that link us today to the evolution of Australia as a nation.

Just as the making of Barangaroo required the removal of the site's industrial heritage, so too has it displaced the long-term residential community within the adjacent area of Millers Point. Long-term residents who have traced their family connections in the area back to the 1800s have been caught in the process of being forcibly relocated out of the area. Millers Point has been predominantly made up of public housing for much of the past century. When it was a home for dockworkers, the area was managed by a state government department called the Maritime Housing Board (MSB), but later this function was transferred to the Department of Housing. The children of waterside workers who grew up in homes managed by the MSB found themselves living in state-subsidised public housing, and many of these residents would remain in their homes for much of their lives, even as the harbour lost its industrial character and was rapidly gentrified. However, as the state government, through its Barangaroo Delivery Authority, was working to redevelop the Barangaroo Precinct, it was also progressively selling this prime waterfront housing stock.

The City of Sydney's long-serving mayor, Clover Moore, has described the process of their forced eviction as 'social cleansing' (Hasham and McKenny, 2014). Many years of campaigning by residents, resident action groups and community supporters have failed to deter the state government from selling 293 public housing dwellings and evicting their residents, some of whom are over seventy years of age and have lived in their properties their whole lives. For those who actively support the residents of Millers Point, Barangaroo has helped to reinforce the spatial divides of the city that separate rich from poor. The removal of the industrial heritage of the precinct is seen to justify the removal of its working-class residents as well. So if Barangaroo helps us to connect with representations of its ecological history, it would also appear to do so at the expense of the area's post-colonial social history.

As this discussion has made clear, there are many competing and highly emotive accounts of Barangaroo as a place, raising many provocations about the nature of its contributions to the life of Sydney. For some, Barangaroo exemplifies why Sydney-siders have so little trust in their government, with its persistently pro-development stance that continues to dislocate the lives of so many, and particularly the most vulnerable. For others, the place points to a new development model capable of reducing the carbon intensity of cities, raising new benchmarks for environmental performance while embracing the site-specific qualities of Sydney's material landscape. For many, this is simply a beautiful new parkland, a place to exercise, a new vista on a glistening harbour.

Arrivals and departures

Esem Projects first began to explore the potential for site-specific audio-visual installations in 2011 while working with the residential community of Millers Point. Once the centre of Sydney's thriving maritime industry, the area is rich in archival documentation, having inspired generations of photographers, film-makers, performers and artists to capture and document changing experiences of life and labour around the industrial harbour.

Working with the support of the National Film and Sound Archive and commissioned through the City of Sydney's annual *Art and About* Festival (2003-ongoing),[8] Esem Projects created a series of site-specific installations called *Unguarded Moments* (2011).[9] In this, we used the mediums of digital projection, sound design and mobile media to reinscribe audio-visual collections back into their environments. The process of designing interfaces that facilitated different kinds of interaction between past and present environments was critical to this work. So, too, were the filmic editing techniques we used, which sought to reposition a character or a moment in time outside their original narrative references, to instead be experienced as a 'trace element' within the environment.

The funding we received from the City of Sydney encouraged us to consider working with the local community in developing this public artwork. For the creative director of *Art and About*, the process of creating public artworks was as much about producing work in the public domain as it was about co-creating with the public. Through a series of workshops with past and present residents of Millers Point, we digitised and recorded personal photographs, oral histories and stories of life growing up in Millers Point. In turn, recordings from contemporary residents were incorporated into the filmic responses exhibited using digital projection within the built environment, alongside the materials we had sourced from public collections. A poor girl eating a sandwich, captured by Australia's first 'mockumentary' producer Rupert Kathner in the 1930s, was slowed down and played on a loop, peering out mournfully from the gentrified shop windows of Walsh Bay; boys playing in a row boat around a damaged ship in the 1940s could be seen, ghostly projected, beneath the docks, laughing back at us.

From this original installation in 2011 our creative work has continued to explore the possibilities of a public art practice that facilitates intimate encounters with those who have come before us and the people who recorded them. We have reimagined the histories of places as archaeologies of recorded action, whereby acts of historical storytelling and interpretation are positioned as deliberately performative and collaborative. Our interest is to facilitate alternate modalities of encounter with different layers of the past, drawing on the mediums of sound design, large- and small-scale projection, print media and experience design. The art of placemaking, within our practice, means enlivening different *times of space* beyond those evident as traces within the built environment today. It is a practice built out of resistance to the notion that time is necessarily linear, and that space is empty until we fill it.

By 2015, when we were commissioned to produce a creative work within the contested landscape of Barangaroo, many of the residents we had collaborated with as part of the *Unguarded Moments* project were in the process of being evicted from their homes. Their stories and the personal photographs they had shared with us during the course of the 2011 installation were central to the story of Barangaroo's maritime history. These residents were without a voice in the story of Barangaroo's future, a site that had deliberately turned its back on its 'ugly' industrial heritage in favour of a pre-colonial, naturalistic past. Protest banners filled the residential streets of Millers Point, but they had no place within the 'welcoming' events taking place over the road at Barangaroo.

As geographer Sibley (1995, p.ix) has reminded us, the human landscape can be read as a landscape of exclusion. Urban revitalisation projects, creating premium, highly palatable spaces, also enact exclusionary practices, often in subtle ways. The residents of Millers Point, living adjacent to the revitalisation of East Darling Harbour, belonged to another era, the era of the working port, when poverty, not plenty, festooned the shoreline. For some the launch of Barangaroo was welcome, but for the residents, this was an act of banishment – just as, when Europeans first arrived here, the people of the Eora nation who had fished the harbour for millennia were also banished.

During our period of creative development leading up to the installation launch, we discovered another narrative of the place somewhat hidden within the historical accounts of East Darling Harbour. The site of Barangaroo Headland had, prior to the containerisation of the working port, been the location of a wharf called 'Dalgetty's Wharf'. It was here, we discovered, that Australians fighting in the two World Wars had embarked to fight in those conflicts, and those who made it home returned to this point in the following years, emaciated and war-weary. A former resident recounted that she had heard a woman singing out on the wharf every time a ship departed, carrying men to war. It was the Maori 'Farewell Song' she had sung, popularised during World War Two. We also learned that Dalgetty's Wharf was the first site of disembarkation for many migrants arriving in Sydney from war-torn Europe. We made contact with people who could share their stories and memories of their first encounter with Australia, who held that moment of first arrival close to their hearts, marking their commencement of a new, better life.

Arrivals and Departures, the title of our installation, acknowledged this migration history, now erased from the landscape, but also contained a reference to the continual forces of inclusion and exclusion, placement and displacement, entailed in the making of a place over time. The design of our installation used shipping containers, a cliché of many placemaking initiatives, but in this instance a direct reference to the history of the site as Australia's largest containerised port. Six containers, each brightly coloured, contained within them a series of films, soundscapes, printed graphics and storytelling opportunities that offered experiential encounters with different narratives of Barangaroo's maritime history. We called each container a 'storybox'.

The use of multiple, brightly coloured storyboxes in the design of the installation contained a reference to the classic arts of memory (Yates, 1966). Originally developed as a series of mnemonic techniques to support the art of rhetoric, this practice involved creating the rooms of a house, in which highly memorable, representational images would be placed as a trigger to provoke recollections. In this installation, each storybox told the life of the precinct from the perspectives of workers, children, recently-arrived immigrants, soldiers departing for war and merchants. One storybox offered a simple meditation on the loneliness of the sea. Another contained a vintage typewriter, along with example letters written by soldiers far from their spouses. Visitors were able to share their own stories, and we found that many people left personal accounts of arriving in Australia as an immigrant.

We deliberately included interviews and documentation provided by the residents of Millers Point, who spoke of their life growing up in the area, swimming in the harbour, avoiding sharks, playing with billycarts. Recordings donated by the Maritime Union of Australia were used to capture working conditions on the docks and the importance of the site to the history of organised labour. Against the intimate interior scale of the containers, we also presented large-scale video projections to create a sense of dramatic scale appropriate to the volume of the space. The Maori *Farewell Song* echoed throughout the cavernous space of The Cutaway.

The project was installed for one month during the Welcome Celebrations and attended by some 80,000 visitors. We ran tours of the project each weekend, and many people asked whether the installations might be made permanent. They told us they had no idea the history of the precinct was so rich. We knew permanence wasn't an option. The site was being redeveloped not to acknowledge and celebrate its maritime history, but to move beyond it, to embrace a vision of Sydney that necessarily departed from the industrial, working-class element of this history.

Conclusion

Arrivals and Departures was a minor intervention within a big, evolving story about Sydney as a city. It was a temporary, creative response to the historical contingencies of place. The installation contributed to the urban revitalisation efforts of the Barangaroo Delivery Authority: it attracted many visitors and caused little controversy. There were no protest banners erected by residents. Moreover, it did nothing to prevent the eviction of Miller's Point's long-term residents. The project's title, as well as its emotional register, was as much about the story of migration, the working conditions of the poor and the loneliness of the sea as it was about displacement, not only of the local residents, but of the many layers of Sydney's industrial history enacted by the arrival of Barangaroo.

The tensions I have described are, I am aware, common to many placemaking contexts where the risks associated with 'pseudo-participation' and a naive romanticism towards the values of place have prompted calls for deeper engagement with the socially-engaged practices that resist neoliberal agendas attached to city culturalisation (Zukin, 2013; Courage, 2017, p.53). If history is told by the victors, the celebratory narratives of places and our enduring, emotional investments in them can, likewise, be easily refashioned to suit the marketisation and privatisation of public space.

Such tensions are no doubt particularly common within creative placemaking projects reliant on funding from developer interests, whether government-owned or private. In places like Australia, and particularly the major cities, it can feel like there are limited alternatives: public funding for community-based arts practice remains low compared to the resources available for urban revitalisation. As Courage (2017, p.77) has put it, such contexts can inadvertently see placemakers complicit in the politics they may be working to subvert or act outside of. The longer-term possibilities for interpretation, community engagement and creativity ultimately depend on the alignment of these practices with the agendas of the site owner, in our case the state-run enterprise known as the Barangaroo Delivery Authority.

How, then, might we respond to these tensions? It seems clear that increased investment in urban revitalisation by entities whose primary driver is footfall can only foster more emphasis on opportunistic engagement. This, in turn, risks seeing the practice of creative placemaking become ever more closely entwined with narratives of developer-led revitalisation. But it is during such times that creative placemaking practice must seek to demonstrate a wider diversity of potentials, including, not least, the creation of inclusive places that speak to the possibilities of spatial justice.

Creative placemaking practitioners are often equipped with tools and resources that can engage a diversity of communities, through collaborative, multi-disciplinary practices and mediums of engagement and expression. Practitioners are able to draw on a range of affective and emotional registers not always present in either more linear accounts of urban history or the celebratory accounts attached to revitalisation agendas. However, to do socially-engaged practice in a way that goes beyond mere tactics or short-term opportunism ultimately depends on resources not commonly available through either arts funding or developer-led investments. A clear priority for creative placemaking policy is to address the limited support available through urban revitalisation mechanisms for sustained, socially-inclusive community engagement.

Some practitioners, such as myself, pull together support for more socially-engaged and/or less commercially-driven practice by knitting together small grants from a range of research, arts and non-profit organisations. However, this approach can lead to a dilution of creatively-engaged practice in favour of more utilitarian goals that ultimately may better be served in other ways. Cultural policy more broadly has experienced this, through the embrace of instrumental policy that has seen arts investments justified for their impacts on the domains of social policy, health policy, economic vitality and the like. Creative placemaking practice can likewise serve a range of diverse agendas, socially-inclusive or otherwise, but the question remains as to whether its identity as a domain of practice should be so instrumentally defined.

For what it is worth, my own attraction to creative placemaking practice lies with its potential for lively encounters with strangers, in ways that affirm and make visible the multiple ways in which people attach and create meaning in places, both past and present. In a world where our experience of place becomes increasingly mediated as not-dissimilar brandscapes, or experienced primarily through the prisms of smallish, glowing rectangular screens, it is the potential to allow space to speak to a multitude of emotive entanglements, memories and enchantments that offers the greatest reward.

Notes

1 Destination NSW, www.destinationnsw.com.au
2 Vivid Sydney, www.vividsydney.com/
3 Esem Projects, http://esemprojects.com
4 Arrivals and Departures video documentation, https://vimeo.com/159049301 and https://www.esemprojects.com/project/arrivals-departures/
5 Barangaroo Delivery Authority, www.barangaroo.com/the-project/barangaroo-delivery-authority/
6 High Line, www.thehighline.org
7 Queen Elisabeth Olympic Park, www.queenelizabetholympicpark.co.uk
8 Art and About Festival, www.artandabout.com.au/
9 Unguarded Moments Project Archive, http://cargocollective.com/unguardedmoments

References

Amin, A. (2006). 'Collective culture and urban public space' in: *Inclusive Cities: Challenges of Urban Diversity*. Woodrow Wilson International the Center for Scholars, the Development Bank of Southern Africa and the CCCB.

Boyer, M. C. (1987). *Dreaming the Rational City: The Myth of American City Planning*. Cambridge: MIT.

Courage, C. (2017). *Arts in Place: The Arts, the Urban and Social Practice*. London: Routledge.

Drew, P. (2015). 'The rise and rise of Barangaroo', *ArchitectureAU*. Available at: http://architectureau.com/articles/the-rise-and-rise-of-barangaroo/ [Accessed: 7 August 2017].

Flannery, T. (2017). 'In praise of sandstone', *New York Review of Books*. New York. Available at: www.nybooks.com/articles/2017/06/22/in-praise-of-sandstone/ [Accessed: 7 August 2017].

Hasham, N. and McKenny, L. (2014). 'Sydney mayor Clover Moore slams NSW state housing sell-off', *Sydney Morning Herald*, 19 March. Available at www.smh.com.au/nsw/sydney-mayor-clover-moore-slams-nsw-state-housing-selloff-20140319-352wf.html [Accessed: 27 August 2017].

Legge, K. (2015). 'How Paul Keating saved Barangaroo headland park on Sydney Harbour', *The Australian*, 3rd October 2015. Available at: www.theaustralian.com.au/life/weekend-australian-magazine/how-paul-keating-saved-barangaroo-headland-park-on-sydney-harbour/news-story/d810af02b77275ad1fcc08e681a81d40 [Accessed: 7 August 2017].

Newman, P. and Kenworthy, J. (1999). *Sustainability and Cities*. Washington: Island Press.

Place Leaders. 2017. *Big Ideas in Place: Conference* 2017. Available at: www.theconnectionrhodes.com.au/event/big-ideas-place-conference-2017. [Accessed: 25 May 2018].

Reinmuth, G. (2012). 'Barangaroo: The loss of trust?', *The Conversation*, 20 November. Available at: http://theconversation.com/barangaroo-the-loss-of-trust-10676. [Accessed: 19 August 2017].

Sibley, D. (1995). *Geographies of Exclusion*. London: Routledge.

Weller, R. (2010) 'Barangaroo – Can it work?', *ArchitectureAU*. Available at http://architectureau.com/articles/can-barangaroo-work/ [Accessed: 19 August 2017].

Wilbur, S. (2015). 'It's about Time: Creative placemaking and performance analytics', *Performance Research* 20.

Yates, F. (1966). *The Art of Memory*. Chicago: University of Chicago Press.

Zukin, S. (2013). 'Whose culture? Whose city?' in: Lin JaM, C. (ed) *The Urban Sociology Reader*. Abingdon: Routledge.

5 A case for human-scale social space in Mumbai

Aditi Nargundkar Pathak

Abstract

This chapter introduces a placemaking initiative that The Urban Vision, a social venture focusing on solutions-driven research, executed in three localities in Mumbai, India. The Urban Vision experimented with the creation of social spaces as a tactic to revive the human scale and social interaction at a neighbourhood level. This was a co-created action research project across three sites that used art as a methodology to investigate the impact of an addition of social spaces. The place activation programme was designed to allow for flexibility of use in an attempt to create ownership amongst the citizens. The projects are discussed as an iterative approach to introducing social spaces or 'pause points' in a busy crowded city. The chapter will present the three projects through an examination of the space characteristics, the processes involved, and post-implementation observations. The chapter will conclude with a reflection on the the necessary adaptations to arts-led interventional placemaking required for a South Asian mega-city, such as Mumbai.

Introduction

In an Indian city the utility of spaces often evolves organically as per citizens' needs, irrespective of the design purpose of those spaces. Children play wherever they find a space, and commerce happens wherever economic opportunity exists. People like flexibility and are averse to rigidity. Alexander (1965) demonstrated this human-focused nature of cities, comparing cities to a 'semi-lattice' in contradistinction to a 'tree', contrasting 'natural cities', those that have evolved spontaneously with 'artificial cities', those that have been designed from the ground up. Alexander contended that artificial cities lack some essential components as they are built with rigidly allocated functions to spaces, advocating instead that cities work best when they mimic the dynamism of the life that they hold.

Mumbai is a dynamic city and characteristically mimics the life of its people. It has grown as an intrinsically organic city and is constantly adapting to rapid population growth. In doing so, it is transforming its organic arrangement by introducing large, planned, rigid places. Mumbai is the second most densely populated city in the world, with 31,700 people per sq. km (UN Habitat for a Better Future, 2015). With the Indian population changing from being predominantly rural to urban in nature and Mumbai being a preferred choice for many because of the economic opportunities it offers, this trend is set to continue, and vertical growth is seen as a feasible solution to this challenge. In every locality/ neighbourhood, there are informally accepted anchor points (Golledge and Rushton, 1976) where people meet, talk or just loiter around. In Indian cities these anchor points have traditionally been a magazine vendor, a tea seller, a street vendor, a popular restaurant, a popular store or sometimes even a bus stop. However, in new townships, the rigidity of the functions within spaces does not encourage such anchor points. 'Old' is rapidly giving way to 'new' and people are losing their affinity to and relationship with their cityscapes. At this inflection point, it has become essential to revive and keep the social fabric of cities like Mumbai intact. One way to achieve this is via the provision of non-intimidating, secular, small social urban spaces alongside large urban amenities. These neutral spaces could hold the key to a more inclusive and well-connected future for Indian cities.

The Urban Vision approach

The Urban Vision,[1] a social venture focusing on solutions-driven research, experimented with creating flexible social spaces or 'pause points' as part of a placemaking programme in Mumbai between 2012 and 2014. The research and implementation phases were led by the author and delivered by The Urban Vision team of urban innovators, architects, artists and urban designers. The Urban Vision's aim was to add social spaces that were left open to interpretation by users, to explore approaches that are more commonly practised in Western contexts; the intent was to evaluate if such spaces could work in the Mumbai setting and what, if any modifications were needed to make a social space successful. The core questions that were intended to be explored were: can such spaces bring legibility and identity to local areas? Can they bring down the enormity of the mega-city to a human scale so that the citizens can relate and interact with them? Can they become inclusive spaces where people of all strata can mingle? Can they act as a catalyst for the revitalisation of their neighbourhoods? And can our group advance any site-specific interventions that enable social cohesion? The Urban Vision team worked with stakeholders – landowners, artists, citizens and supporters – to design and deliver social spaces in three localities in Mumbai. Two of the three spaces were built with citizens' participation in already-existing localities and the third space was in a new planned residential township. The three spaces represent different scenarios of the city even though all three sites were implemented on private land owned by private developers. In Mumbai, permission to develop private land is easier to acquire and funding for is easier to secure from the owners than via public sources. Developers saw value in creating the pause points to enhance the ambience of their projects. The spaces were located on the main road of the neighbourhood for easy access, and with adjacent bus links to enhance footfall they acted as an extension of the street for pedestrians. Visual art was used as method of intervention across all three sites. In the recent past, only architecture and infrastructure had been used to create a sense of identity in Mumbai. Thus, these projects were an experiment to see if art-led temporary interventions could impact that status quo. Social spaces as pause points and accessible spaces were created to fulfil some of the perceived needs of social life. Art was used as a way of making the space inviting and non-intimidating by aiming to harness feelings of happiness and relaxation, stimulating the pleasure centres of the brain which make people feel happy and relaxed (Zeki, 1999).

Project context

The three spaces that were chosen for the interventions were located in three neighbourhoods of Mumbai. Although at the city level all these spaces were small, they had the potential to showcase the impact of interventions through an analysis of their ability to mitigate the monotony of new urban environments. These spaces could potentially provide a hint to a better future for Indian cities having inclusive small social spaces. The owners of the private land on which the three projects were built were also keen to support an initiative where improving walkability and scaling the city down to human level was a primary objective.

The creation of two temporary and one permanent space employed two different approaches to place activation: first, temporary interventions that explore PPS's (2011) Lighter-Quicker-Cheaper (LQC) approach to placemaking; and second, the use of permanent art. These approaches, combined with site observations and a comparative technique, enabled The Urban Vision to explore whether methods developed within Western contexts could be transferred to India. Project 1 used tactical urbanism in a popular neighbourhood. Project 2 used mural and street art to mitigate a building site. Project 3 used permanent art and place programming to energise a newly built township.

Project 1: Temporary social space in Powai

This site was located in Powai, a popular neighbourhood in North Mumbai. The space consisted of a planned development by a single developer and was located adjacent to a natural lake. The area has a mixed resident demographic and a transient professional population. The locality has commercial office spaces, residential buildings, shops, restaurants, a hospital, schools and one of the premier educational institutions of the country. All these amenities are located in a small area of approximately 1.5 sq. km and the facilities have made this neighbourhood a desired location in which to live and work. The mutual proximity of amenities, workspaces and apartment blocks made the site for this social space a fitting location for a pause point, and the designers sought to engage various stakeholders in building a social space that could mitigate the enormity of the uniform high-rise buildings dominating the neighbourhood. It was thought that incorporating public art in the form of murals and sculptures would change the area's aesthetic appeal, and observations on the extent to which this changed aesthetic was accepted by locals would form the findings of the research project.

Project 2: Temporary social space in Borivali

Borivali is a suburb to the north of Powai, originally developed as an industrial neighbourhood. Here, real estate prices have experienced steady growth each year, and premium land prices and residential demand has led to conversions of industrial land to residential spaces (Housing.com, 2017). This development, and the project site, was on the Western Express Highway, the main north–south arterial road of Mumbai. It is estimated that 62,191 vehicles pass through this road daily (Maharashtra State Road Transport Corporation Limited, 2016), making the site a prominent location. The neighbourhood had residential apartments, a hospital, a mall and a national park close to the site. Prior to the intervention the site was cordoned off with plain grey galvanised iron sheets and was visually lacklustre. The project was supported by volunteers on a *pro bono* basis and the landowner provided funding for materials, basic civil works and an honorarium for artists.

Project 3: Permanent space in new township, Thana

The Urban Vision developed and implemented an art and place activation programme to enable civic engagement with a new planned residential township in Thana. Thana neighbours Mumbai and is a district in the Mumbai Metropolitan Region. The location of this planned development is approximately 34 km from Mumbai, developed as a residential township with some city amenities including offices, shops, a school, fire stations, restaurants, gyms and health clubs. The challenge of this new township is to provide a high-quality lifestyle for residents in order to attract new buyers. The management and developers of such townships are constantly applying added-value approaches to projects. New developments of this scale bring people from different backgrounds together and there is a general absence of pre-existing social connections within such communities. At the time of the project, therefore, the township was still a construction site, and the installations were to take three to six months to complete. Four art organisations were selected from an open tender to complete seven installations.

Methods and methodology

The temporary spaces employed an experimental methodology using three key methods: a hackathon, a social media campaign and co-creation. Hackathons are generally a coding festival used by information technology professionals to rapidly develop software products. City design professionals and citizens have been using hackathon methods to organise short (one- or two-day) brainstorming sessions to devise user-friendly solutions for city planning and urban design. Here, citizens, artists and urban planners brainstormed to come up with design solutions for the space. A social media campaign via Facebook and Twitter (both widely used mediums in India for information outreach) was used to invite people to the hackathon and later to take part in the delivery phase. The temporary spaces were built via co-creation with local citizens. The creation of the permanent spaces employed a more traditional approach involving, firstly, the identification of the site, then the development of the budget, the commissioning of art work, the site preparation and installation, and lastly, communication and publicity.

Temporary spaces in Borivali and Powai

The spaces at Powai and Borivali were to be built as temporary spaces using PPS's LQC approach and terminology (2011) as advocated for by Project for Public Spaces (Project for Public Spaces, 2011). The temporary spaces created with this methodology are easy to build and have a relatively lower cost, and thus were considered to be best suited for the context as they are realisable even with a limited budget and timeframe. The LQC approach further allowed the use of straightforward and practical solutions where the community could be involved from planning to execution. Co- creation with citizens meant completing works in the early morning and late evening within a short time span. These first projects, executed in four days, were an important step in showing proof of concept and how much of an impact could be made by a quick transformation of part of the city.

In Powai, commercial activity was not allowed to be conducted at the space, so a social anchor point such as a tea stall could not be provided – street art and interactive sculptures were therefore used instead. The purpose of art in this project was also to give a social and visual cue to passers-by that they are welcome to use this space, and so to achieve this the commissioned artists chose vibrant and joyful colours as an initial design feature. In the visually demanding landscape of Mumbai, art needs to create a striking impact to draw attention to space and to make the different functional qualities of space recognisable, such as the implied invitation to linger or sit, for example. An existing dilapidated structure was repurposed to create a Museum of Street Art. Street art groups Visual Disobedience[2] and ST+Art India[3] were invited to paint the walls of the museum and interactive sculptures made by artists were installed, such as *FOSSILSS*, a sculpture of a family with a dog created from recycled material. People could shake hands with the father, mother and the child, as the hands were movable, and also sit on the dog. A three-dimensional piece of floor art was also incorporated on the pavement adjoining the social space.

At Borivali, the intention was to carve out a temporary social space that would be open to people for the duration the construction phasing and which showcased how construction sites could be used for meanwhile activity. The site was located behind a city bus stop; this location was selected in order to integrate the social space into the street and make it accessible and non-restrictive. Two types of art interventions were undertaken here. Firstly, a social space was created by artists and residents. Recycled tires were converted into colourful seats and installed in an amphitheatre formation, enabling a dynamic interpretation as seating, a performing space or a children's play area. The walls of this social space were inscribed with visions for the future of Mumbai tweeted on social media by citizens and some quotes on urbanism best practices. Aklaq Ahemad, a 'letter artist', was commissioned for this project. Secondly, a mural was created by the graffiti/street art groups Visual Disobedience and ST+ Art India. Five murals were made from 60 x 16 ft galvanised iron sheets found on the construction site and painted offering an artists' interpretation of visions for a future Mumbai.

The subject of the art work emerged from a hackathon. The artists painted their interpretation of the vision based on their individual styles, with locals from the neighbourhood painting parts of the murals. Social media engagement with was an important aspect in discovering and then inscribing citizens' wishes on the walls of the social plaza, for example, whereby messages such as #Needmoresuchplacesto-displayourtalent, #moretimetohangoutwithfriends and #MumbaiNeedsmorepedes-trianfriendlyStreets were included.

The Urban Vision had organised a design hackathon earlier, in 2014, when residents, artists, planners, architects and students were invited to collaborate on a brainstorming session to design the art-led small social space in Powai and Borivali. This was a five-hour session that produced solutions for the two temporary sites in Powai and Borivali. In an attempt to involve the larger city, the chosen solutions were uploaded to social media. Social and print media were used to engage the city and invite citizens to help with their execution. Twitter was used to follow the progress of the build and featured film interviews with residents. Events such as artist talks, a drum circle and storytelling were organised to encourage citizens' participation. The project was positively reported by print media and was covered by national television and both English and Marathi newspapers (Cathcart-Keays, 2014; Doshi, 2014).

Permanent art and place activation

For the Thana development, The Urban Vision utilised public art and place activation, or place programming, as a strategy to increase the number of social spaces, create pause points and encourage greater sociability among locals. Permanent art installations were used to add character and identity to the already-designed green spaces and to counter the homogenised yet unintelligible architectural aesthetic designed for monetary purposes. The objective of using curated permanent art commissioned from multiple artists was to add legibility to the existing community spaces, giving people landmarks or anchor points and creating visual associations within the township. Place activation was used to promote citizen involvement in the common spaces and foster relationship building among fellow citizens. Citizen engagement in this context was not used as a planning term but as a method of encouraging residents to actively engage with both their built environment and their fellow neighbours. This engagement was achieved by conducting community programming, forming interest groups and promoting the identity of the full township within the city as a destination for new experiences. Place activation was used as a non-physical placemaking strategy to engage the community in order to 'activate' the spaces they had moved into and in turn create community ownership.

PERMANENT ART INSTALLATIONS

Local citizens were not involved in the planning or delivery of these art works; instead, the developer created an overarching vision through a series of exercises undertaken with The Urban Vision. As this was not a temporary approach, the installation locations were selected after observing how citizens used the spaces and understanding the development design. It was felt to be important to commission art that was inviting and accessible, that could be touched and played with and that could be integrated into the whole development, merging with the language of the township. A further commissioning criterion was the overarching sustainability theme of utilising waste material within some of the sculptures and conceptually showcasing green living. Seven locations, at regular ten- to fifteen-minute walking intervals, were selected for these pieces.

FUNCTIONAL ART INSTALLATIONS

Floor games such as hopscotch and mazes were installed at bus stops and 'artistic seating' was installed in the community greens along with indigenous nature-inspired lighting on the main connecting road of the township. In-situ seating around the street lighting was also installed to create a social space. A 20 ft globe made from recycled iron and stainless steel was installed at the main entrance of the township. As the largest installation it needed to be located in a landmark location and a space was chosen between the development sales office and one of the schools. This location conveyed the township's commitment to green initiatives and used bicycle dynamos to light up a map of India when peddled, encouraging an awareness of green energy. Eco Twirls, aluminium fans that worked as windmills, were installed on three walls on a sub-station wall in front of children's play area and on the way to the grocery store and health club. These were made into a competitive game for toddlers and became a loved play space that was consistently used by the children and adults.

PLACE ACTIVATION FOR CITIZEN ENGAGEMENT

Programmes and events were designed to encourage citizens to form social bonds within the community and with an aim for citizens to, in due course, take over and conduct their own programmes. The events were categorised into Partner Events – cultural and social events hosted within various sites in collaboration with partner organisations – and Community Events – small-scale events developed and conducted by citizens. The intention of the programmes was to create an environment for new social networks to form, create a sense of belonging and patronage of a place and establish a feeling of security and safety in the community through processes of trust building. The programming content was based on citizen surveys and stakeholder workshops and included the Rangoli festival, a lantern festival, photography, a capoeira performance, dance and aerodynamics workshops and urban farming.

Findings

On completion, observations were made to evaluate the spaces as per the core questions that The Urban Vision had set out to explore. These observations were noted through photography and observational data collection. Interviews of users were also conducted from 6 am to 8 am and at evenings from 5 pm to 8 pm. The temporary sites were observed twelve times in the first year and six times in second year for all the selected times. The permanent art installations were observed over a period of eighteen months after installations. Observations focussed on recording the number of people per hour using the space at predetermined times and different days of the week, the condition of deterioration and vandalism and changes in physical conditions.

LQC as a strategy

It was found necessary to tailor the LQC approach according to the site locations. At the Powai site a parking area was dilapidated and required work from the land owners to make it accessible. Similarly at Borivali, the developer did the base work of carving out the space and painting the primer coat on the murals.

Social spaces as pause points

The space in Powai achieved its objective of creating a pause point. The space was most often used in early mornings and late evenings as the micro-climate was cooler, and citizens were observed using the space for a break during their morning walk. Young people used the space the evenings and a greater number of women and children were seen spending time in the space and its adjacent street, indicating that the space was perceived as safe. The position of seating made users face the street adjacent to the space instead of the art and while people chose to look at the street, art functioned as a visual cue that the space was safe and accessible. Those observed were from all the same strata of society: users interviewed included rickshaw drivers, professional drivers, domestic helps, nannies, a grandmother, homemakers from affluent families, shopkeepers from nearby shops and security guards.

Meanwhile activity and social space

At Borivali the value of opening up an under-construction site to the city attracted attention. The social space and the art murals made the space visually noticeable, leading to other developer organisations approaching The Urban vision team to implement similar murals. Developers have started considering this an innovative approach to establishing community goodwill and their brand presence around the ongoing project sites. The vibrancy of the wall mural was successful in mitigating the harshness of the construction, as this investment converted a derelict-looking space to one that was well taken care of and valued. During the two years the social space was in use there was not much deterioration (as maintenance was discharged by the developer) and the space was in use until its removal after two years. A shade device would have been useful as it was observed that the space was unused during the summer or when it rained, or in the afternoons generally.

Lack of maintenance and vandalism

At Powai, it was observed that the social space began to deteriorate, with parts of the sculptures missing and not replaced as there was no clear ownership established during commissioning. The citizens groups involved expected the developer to take the lead in maintenance and vice versa. The direct impact of the lack of maintenance was seen in the reduced use of the space by women and children in the second year. At Borivali, the developer experienced vandalism on the site during the two-year project and subsequently employed a night-time security guard. An activity anchor such as a vendor would have helped to establish a sense of local ownership and thus avoided a guard being present there.

Permanent art for pause points

This project was a pilot project to see if art could create social spaces and enhance social interaction. The locations of each installation became social spaces, with people interacting with their neighbours around the installations, turning an unnoticed space into a pause point for the residents during their day-to-day activities. It was observed that people were recognising the art pieces as indicators and using them as anchor points. The new commissions given to the artists exemplified an emerging trend to include art in new developments as part of a modern approach to create social spaces and to add uniqueness to the new townships.

Place activation for increasing sociability

The programmes selected for activating the spaces were well received. A few citizens' clubs were formed, which continued the programming. City management,

along with some citizen representatives, has also continued to manage the programmes and the partners involved have secured further projects.

Conclusion

In the older parts of Indian cities it is common to see a bookstore next to a
residence, which is in turn next to a police station that is in front of a hospital.
These spaces formed organically to support local life. This organic arrangement
made sure that cities were walkable and convenient for social life. However,
newer developments are rigid, with separated service functions, which makes
social interaction challenging. Small or human-scale social spaces, or 'pause
points', were The Urban Vision's modest attempt to mitigate this rigidity and
allow the creation of spaces without any function within the city. Visual art was
found to be effective in signalling that such spaces are being taken care of and
offer security and subtle cues that hint towards the inclusion of all citizens. In the
examples given here it was found that the LQC approach, while a great tactic for
pilot projects, should be used for a limited time and that public art needs to
evolve to include innovations that help in building great cities. In these projects
only visual arts were used but creative placemaking is a much wider discourse
that can be ventured into.

Co-creation did not help in creating ownership as people volunteered for a day
and then did not take part in the ongoing maintenance. The assignment of spaces
to citizen groups, a delegation of responsibilities along with activities like co-
creation, is needed to create self-sustaining social spaces throughout the city. At
the time of writing, The Urban Vision is working with politically elected
neighbourhood heads to ascertain if they can mobilise businesses and residents'
associations to maintain such spaces, and has successfully completed a small
number of art-led projects that are maintained by the private sector. Some
private-sector real-estate developers have used similar art-led interventions to
promote their projects and gain community goodwill for upcoming projects. The
Urban Vision's project findings have been instrumental in guiding the design
principles for the future projects undertaken.

All the above projects that have been discussed were The Urban Vision's
attempt to build social spaces or pause points in a city and study the results.
There is no single or simple recommendation that can be drawn from all the
experiences described; however, many more opportunities can be found. Coming
back to the questions asked above of the projects at their inception, it is to be
noted here that spaces in Mumbai work differently to their Western counterparts.
Indian cities need spaces that their citizens can relate to. While we agree that art
and place activation as strategies may not be the answer all the time, nevertheless
it is a secular medium, which can be appreciated by all Indian citizens, across a
mix of many classes, cultures, languages and religions. There is a fear that, by
using global interventions like creative placemaking, we may alienate and
exclude already-marginalised citizens. It is therefore essential that artists and
space planners use tactics which are seen as non-intimidating and bring pride to

the community they are placed in. The social media campaign and live Twitter feeds The Urban Vision used documented that people like to visit areas that offer innovative activities and experiences.[4] During the execution of the social spaces, all sections of society came together and provided sweat equity. It is important that the authorities enable and support initiatives that allow citizens to work towards making their urban environment better. Creative placemaking in Indian cities can be used as an ice breaker to make cities more inclusive by encouraging residents to work towards a common goal.

Notes

1 the urban vision, www.theurbanvision.com/blogs/
2 visual disobedience, https://www.facebook.com/VisualDisobedience/
3 st+art india, https://st-artindia.org
4 @theurbanvision #remakebby, https://twitter.com/theurbanvision?lang=en

References

Alexander, C. (1965). 'A city is not a tree', *Architectural Forum*, 22(1).
Cathcart-Keays, A. (2014). 'Can Urban Vision turn Mumbai's unloved plots into precious spaces?', *The Guardian*, 18th November. Available at www.theguardian.com/cities/2014/nov/18/can-urban-vision-turn-mumbais-unloved-plots-into-precious-spaces [Accessed: 25 February 2018].
CitiSpace-Citizens Forum for Protection of Public Spaces. (2010). *Breathing Space: A Fact File of 600 Reserved Public Open Spaces in Greater Mumbai*. Mumbai: CitiSpace.
Doshi, R. (2014). 'Citizens to be roped in for revamping public spaces in Mumbai', *Hindustan Times*, 20th November. Available at: www.hindustantimes.com/mumbai/citizens-to-be-roped-in-for-revamping-public-spaces-in-mumbai/story-kN5fTriTvifXSxJqRLeVyN.html [Accessed: 25 February 2018].
Dwivedi, S. and Mehrotra, R. (2001). *Bombay: The Cities Within*. Mumbai: Eminence Designs Pvt. Ltd.
Gaikwad, S. (2016). 'Developers will get to run townships in Maharashtra for 10 years', *Hindustan Times*. [Online] Available at: www.hindustantimes.com/mumbai-news/developers-will-get-to-run-townships-in-maharashtra-for-10-years/story-b2yBaYe0kPNQykCo88LgbN.html [Accessed: 23 October 2017].
Golledge, R. G. and Rushton, G. (1976). *Spatial Choice and Spatial Behavior: Geographic Essays on the Analysis of Preferences and Perceptions*. Columbus: Ohio State University Press.
Harvey, D. (2003). 'The right to the city', *International Journal of Urban and Regional Research*, 27(4).
Housing.com. (2017). *Borivali East Market Overview*. [Online]. Available at: https://housing.com/news/borivali-east-property-market-overview/. [Accessed: 1 September 2017].
Iese, B. S. (2013). 'Quality of life: Everyone wants it, but what is it?', *Forbes*. [Online] Available at: www.forbes.com/sites/iese/2013/09/04/quality-of-life-everyone-wants-it-but-what-is-it/'"\l"2c8a2b3c635d [Accessed: 23 October 2017].
Jamwal, N. (2006). 'Mumbai may lose its open spaces', *Down To Earth*. [Online] Available at: www.downtoearth.org.in/news/mumbai-may-lose-its-open-spaces-8708 [Accessed: 27 July 2015].

Jog, S. (2015). 'No new taxes in BMC's Rs 33,514.15 cr budget for 2015-16', *Business Standard Economy*. [Online] Available at: www.business-standard.com/article/economy-policy/no-new-taxes-in-bmc-s-rs-33-514-15-cr-budget-for-2015-16-business-standard-news-115020401359_1.html [Accessed: 3 June 2015].

Maharashtra State Road Transport Corporation Limited. (2016). Schedule F - Monthly toll collection statement for Mumbai Entry points, Corridor: Western Express Highway. [Online]. Available at: www.msrdc.org/Site/Upload/GR/MEP%20Daily%20Traffic%20Data%201-8-2016%20to%205-8-2016.pdf. [Accessed: 11 September 2017].

Partners for Livable Communities. *What Is Livability?* [Online]. Available at: http://livable.org/about-us/what-is-livability. [Accessed: 25 June 2015].

Philip Oswalt, K. O. M. (2013). *Urban Catalyst-The Power of Temporary Use*. Berlin: DOM Publishers.

Project for Public Spaces. (2011). *The lighter, quicker, cheaper transformation of public spaces*. [Online]. Available at: www.pps.org/reference/lighter-quicker-cheaper [Accessed: 1 June 2017].

Shanker, S. (2008). 'Mumbai pedestrians can walk safe in the sky' in *The Hindu Business Line*. [Online] Available at: www.thehindubusinessline.com/todays-paper/tp-investment world/article1114744.ece [Accessed: 15 July 2017].

Silberg, S., Lorah, K., Disbrow, R., Muessig, A., and Naparstak, A. (2013). *Places in the Making: How Placemaking Builds Places and Communities*. Boston: Massachusetts Institute of Technology.

Thakur, V. K. (1981). *Urbanisation in Ancient India*. Delhi: Abhinav Publications.

The Economist Intelligence Unit. (2012). 'Best cities ranking and report: A special report from the Economic Intelligence Unit' in *The Economist*. Available at: www.eiu.com/public/topical_report.aspx?campaignid=BestCity2012. [Accessed: 6 July 2015].

Tindall, G. (1992). *City of Gold: The Biography of Bombay*. New Delhi: Penguin Group.

Tucker, A. (2012). 'How does the brain process art?' in *Smithsonian Magazine*. [Online] Available at: www.smithsonianmag.com/science-nature/how-does-the-brain-process-art-80541420. [Accessed: 6 July 2017].

UN Habitat for a Better Future. (2015). *Urban Data*. [Online] Available at: http://urbandata.unhabitat.org/explore-data/?cities=6145andcountries=INandindicators=slum_proportion_living_urban,population,urban_agglomeration_population_density,urban_population_cities,hiv_prevalence_15_to_49_year. [Accessed: 23 October 2017].

Zeki, S. (1999). 'Art and the brain', *Journal of Consciousness Studies*, 6(6–7).

Section 3

Scalable Ecologies

6 A rural case

Beyond creative placemaking

Margo Handwerker

Abstract

This chapter assesses how creative placemaking—which relies largely on the assumption that the quality of a place is built using 'cultural' expertise—is impacting public arts in the United States. As both a researcher and a member of the M12 arts collective, Handwerker provides reflections that are critically and historically-driven, but also informed by practice. Aided by comments from M12's founder Richard Saxton, Handwerker addresses M12's determination to create place-based work that is informed by the realities and complexities of rural life in the United States, but without the pressure to produce 'solutions'—or to be the solution—to the problems these communities sometimes face. Private and public entities, she argues with others, have used creative placemaking to rationalize the siphoning of already limited funds from the arts to social services and private development.

Artists aren't always activists

In their 2010 White Paper for the National Endowment for the Arts, Markusen and Gadwa describe creative placemaking as follows:

> In creative placemaking, partners from public, private, non-profit, and community sectors strategically shape the physical and social character of a neighborhood, town, city, or region around arts and cultural activities. Creative placemaking animates public and private spaces, rejuvenates structures and streetscapes, improves local business viability and public safety, and brings diverse people together to celebrate, inspire, and be inspired.
>
> (Markusen and Gadwa, 2010, p.3)

The language of creative placemaking is widely criticized among public arts professionals as an agent of gentrification. Those of us quick to criticize this language do so with the following quandaries in mind: How is a community's interest determined and by whom? How is improvement – a relative term – determined? How do we judge transformation as successful or unsuccessful?

How do we measure whether or not character and quality are shaped? How does framing the field in these terms impact the artists, the places, and the funders who participate? How are the missions and methods of such disparate participants impacting our field? The language of creative placemaking, despite such reservations, has begun to dominate the cultural atmosphere in which public arts professionals are working, particularly within rural areas. This atmosphere is especially salient to me as I participate in the field both as a researcher, where I have explored the role of art as a tool and of artists as service providers, and as an active member of the M12 group,[1] a collective that makes art work attentive to the character of rural lives and rural landscapes today.

Artist Claes Oldenburg (1961) once remarked that he was 'for an art ... that does something other than sit on its ass in a museum.' While it is true that not all art work belongs in a museum, Oldenburg's aphorism implies that art in museums – and presumably outside of them too – is not already doing something. What does such an expectation presume about what art should be doing instead, and who says so? What might new demands on art replace and, most importantly, where are these demands coming from? Motivated by these questions, many of us engaging in site-responsive practices distance ourselves from the category of 'creative placemaking' for this simple reason: We resist an assumption implied by the phrase, that the quality of a place is actively 'made' by those of us in the creative sector. Art may indeed animate, rejuvenate, and even improve the qualities of a place, to use Markusen and Gadwa's language, but it is too often the case that we take culture to be a commodity that arts professionals are uniquely equipped to harness or bring about, to produce and then to trade.

In a 2016 interview, art critic Lucy Lippard asked artist Richard Saxton, M12's founder: 'How do you feel the notion of Social Sculpture has been expanded in the last couple of decades?' Saxton's response reinforces the difficulty that many of us have with the way that art has been instrumentalized in this way:

> I like that you use the term Social Sculpture, too many people don't use that term. Instead it's 'socially-engaged art' or 'social practice,' 'community art,' or my absolute most disliked phrase 'creative placemaking.' My creative interests led me to work with the Center for Social Sculpture in Minneapolis (now defunct) early on. They were dedicated to continuing Joseph Beuys's 7000 Oaks project and promoting the important ideas behind social sculpture. I think those ideas are actually very different than forms of social art from recent years. I always dug the spiritual part of social sculpture and the embracing of creativity, its boundlessness, and the connections to indigenous, generational thinking. It's hard to answer this question because in some ways our practice is firmly planted in the ideas of social sculpture, but when I look around at other work that gets labeled 'social' today, I don't identify with a lot of it. The thing that's happening now is basically corporate takeover and real estate investors changing communities in the name of art and 'creativity' – the 'creative class' is a disguise for power. I return to Beuys often: 'Only on condition of a radical widening of definition will it be

possible for art and activities related to art to provide evidence that art is now the only evolutionary-revolutionary power.' That's the opening line from *I am searching for field character* [1973]. I think we're still far away from that idea, and I guess I don't really want to speak for social sculpture or social art today because it seems pretty convoluted. I see a lot of things that look like real estate investment and corporate gimmickry. I see a lot of social narcissism. I'm hopeful that the work we do counter balances some of that.

(Saxton, quoted in Lippard, 2016)

It seems true that the collaborative, resourceful ethos characteristic of the public arts today has produced lines of inquiry that seem overly presumptuous in their use of the term 'social.' Even in the rural case, idyllic portrayals of the barn raising and utopic visions for how artists might brighten blighted towns raise criticism from those who suspect that generalizations about whole 'communities' do not account for difference and, therefore, undermine attempts to address the specific qualities or needs of a given region or population. The latter criticism stems from a 2004 essay by Bishop, in which it is argued that we mollify the politics of relational practices by interpreting a 'community' generally, by aestheticizing it. Indeed, policy makers and private corporations have since adopted a 'social practice,' implementing the instrumental quality of this aesthetic. As Saxton remarks above, the result is a kind of gimmickry, one that disguises power.

Couching social art work as social entrepreneurship risks reinforcing the notion that social welfare is sufficiently provided by individuals, private and third sector initiatives instead of by the government. This is not a position that M12 seeks to reinforce, and so we resist the idea that we as artists 'strategically shape' the communities in and with which we work. Such an emphasis on deliverables reinforces the notion that art should quantify its worth. On the role that activism might play in M12's work, Saxton states the following in the same interview with Lippard when she asks in the same interview: 'How much change can art facilitate?' Again, I quote Saxton at length:

I'm not sure I know how to quantify change. I think in some ways change and how art can facilitate change, is tied to the people and place you work with. For us, we've worked in a number of different communities and I think all of them have experienced change in some way through the projects we've initiated. But that said, change is something we think of as being constant – rural communities are in so much flux these days – so good change, or bad change, or lasting change, or systematic change, that might be something to explore further. I guess we hope that we're making some kind of positive change. We simply define that as moving towards a deeper and more present experience with the places we live and work in. Too often it seems that as a larger social being we're moving away from a collective awareness, and away from the small and intimate, poetic, and away from being a supportive and sustainable species. I think in our projects we are trying to get closer to an idea of elemental awareness that exists outside of the city, both for ourselves and for those who engage with our work. We see our practice as a series of connections, much like an aesthetic network or terminal with many ideas, people, and experiences interfacing. Perhaps it's difficult to see the change on the micro level, but once zoomed out, the energy that enters the main terminal becomes something to learn from. If our work can help steer us towards collective awareness and connection, then in the long run we're probably getting closer to change.

(Saxton, quoted in Lippard, 2016)

The connectivity to which Saxton refers is in keeping with his reference to Beuys about widening the definitions of art. Beuys (1973/1990, p.21) described his idea for Social Sculpture/Social Architecture this way: 'This most modern art discipline – Social Sculpture/Social Architecture – will only reach fruition when every living person becomes a creator, a sculptor, or architect of the social organism.' Producing such a social organism – Saxton's own reference to *systematic* change – demands a systems aesthetic, an aesthetic based on interdependence, an aesthetic long undervalued in the history of contemporary art making.

Systems art, a phrase popularized by art historian Jack Burnham (1968) in the late 1960s, is a way of characterizing art's production relative to governing systems, whether those systems are ecological, economic, political, social, and so forth. Burnham and many of his contemporaries make the analogy between systems art and ritual, ritual being the carrying out of an otherwise everyday action in a prescribed way. Lippard (1980, p.364) provides a helpful audit of ritual's significance within art discourse, writing in 1980 about the role of inclusivity in Feminist art:

The popularity of the notion of ritual indicates a nostalgia for times when art had daily significance. However, good ritual art is not a matter of wishful fantasy, of skimming a few alien cultures for an exotic set of images. Useful as they may be as talismans for self-development, these images are only containers. They become ritual in the true sense only when they are filled by a communal impulse that connects the past (the last time we performed this act) and the present (the ritual we are performing now) and the future (will we ever perform it again?). When a ritual doesn't work, it becomes a self-conscious act, an exclusive object involving only the performer. When it does work, it leaves the viewer with a need to do or to participate in this act, or in something similar, again. (Here ritual art becomes propaganda in the good sense – that of spreading the word.) Only in repetition does an isolated act become ritualised, and this is where community comes in.

(Lippard, 1980, p.364)

Rituals can be personal, intimate actions. However, as Lippard suggests, rituals have greater significance when they involve others – when they are carried out in the multiple. Her remarks were written well before the 'social practice' arts emerged, and yet they characterize that movement's underlying principles as well: Social sculpture is *sacra* in the anthropological sense insofar as it can reinforce such normative social beliefs as interdependence and sustainability – that collective, elemental awareness to which Saxton refers. M12 has long created sites and circumstances for gathering and connectivity – a description of 'creative placemaking' if there was one. What follows are some examples of M12's own alternatives to creative placemaking as the phrase is now more commonly understood. They are spaces we have made, places we have created, but without the burden of articulating our *strategies, accomplishments*, or *long-term impacts*. Why not just 'sit,' to return to Oldenburg's adage, without the pressure of *outcomes*?

M12: rendering, not shaping

Last Chance Press (2015–present) is an ongoing initiative that invites experts from a variety of backgrounds and fields to produce small books and audio recordings on a range of themes pertinent to art and rural studies. Each pocket-sized volume explores the changing realities of rural landscapes and communities around the world – presenting an array of curated notes and research ephemera with images, poetry, and more traditional visual and written works. The *Star Route 1* (2018) volume, our most recent, includes photographs, interviews, and historical accounts of regional post offices throughout the Plains of Colorado. The book's editor, Mary (Welcome) Rothlisberger, describes these post offices as 'a place where intimacy and institution collide,' serving as 'the social fulcrum (both publicly and privately) in remote communities.' Last year we produced *This Road Leads to Nowhere: Pierre Punk* (2016),[2] a collection of images and testimonials from an underground music scene that flourished in small-town South Dakota from the early 1990s through to 2010. The volume, which draws on the do-it-yourself punk aesthetic, gathers up personal accounts and ephemera from Pierre community members, musicians who traveled through and played in the town, and kindred spirits from similar scenes further afield. *An Equine Anthology* (2015),[3] the first book produced, is a collection of responses to equine culture from the American Southwest and elsewhere. Recipes for using horsemeat are juxtaposed with histories of rodeo culture, remarks from animal welfare activists, and photographs and paintings by contemporary artists in the West. Each book – these are only three of five published to date – is accompanied by a seven-inch vinyl record that acts as a sonic counterpart to its book. Typically, side A consists of a new or previously unreleased song inspired by the contents of the book, while the B sides offer more experimental soundscapes and spoken word performances. Together, the five pairs produced since the project's inception make up the Press's Center Pivot Series. Like a center pivot irrigates a field, the series provides for an alternative space for productive rural inquiry and creative collaboration between M12 and invited contributors.

In 2015, the pages of *An Equine Anthology* were installed at the Santa Fe Art Institute, alongside M12's *The Breaking Ring* (Figure 6.1),[4] a sculpture and installation commissioned by the Institute as part of its Food Justice initiative and residency. The 24 ft breaking ring (also referred to as a gentling ring or round pen) is made from regional hand-peeled Aspen logs. Its form is a reference to wild horse culture in the American West and to the built environment there. Intended to host a diverse selection of public programs throughout the duration of each installation, the enclosure is also a metaphor for the physical and intellectual spaces required for grappling with those ideas and relationships that are the most challenging. Despite the violence of the name, 'breaking' occurs less through force than through the building of trust. And, yet, this reference to violence is useful in the case of *The Breaking Ring*, where conversations about the economic and social realities of rural lives and landscapes, can at first be heated; resolved only through patience and mutual respect, by finding connection, on common ground.

Both the books and *The Breaking Ring* are 'placemaking' to be sure, but they are not bound to any one place – they are re-contextualized with each new site, whether found in another museum installation, an open field, or a pair of hands. They are site-responsive and yet site-less – these qualities are not mutually

Figure 6.1 M12 Studio (2016). *The Breaking Ring*. On view at the Center for Contemporary Arts Santa Fe, New Mexico. Installation, size variable. Image courtesy of M12 Studio.

exclusive in the rural case where attention to both the local and to the non-local is crucial. Individual rural communities span large distances and those who practice in and with these communities rarely network in one place the way that other artists converge in major art centers. Online platforms and convenings are increasingly more common, so that groups like M12 can maintain their place-based character without losing access to a more extensive dialogue. Itinerancy is one among many rural conditions. Rural neighbors are separated by as many as 75 miles, and so driving, a lot of driving, is common. This reality is reinforced by the irony that some of our group's more successful 'creative placemaking' grants were awarded for projects we might also characterize as mobile (*Gran's University*, Kultivator and M12, (2012)[5]) or temporary (*Turecek Murals*, Jetsonor-ama and M12 (2011)) – qualities that seem to move further away from the objective of creative placemaking's funders,' to improve the livability of a place through infrastructure that is both contextual and long-standing.

M12's practice is a connective one, weaving a range of expertise to realize a range of creative outcomes, whether a workshop, a book, a record, radio programming, or a physical space. Motivated by care and concern, individuals with varying kinds of knowledge, whether artistic, regional, scientific, etc., come together to create something of mutual interest and shared effort. Conceptualizing and executing projects through collaboration with various stakeholders is key to our connective aesthetic, and forging these relationships takes time. It also takes space. In 1980, artist Robert Morris (p.98) wrote about the problem, even then, of finding such resources for artists: 'The key that fits the lock to the bank is "land reclamation." Art functioning as land reclamation has a potential sponsorship in millions of dollars and a possible location over hundreds of thousands of acres throughout the country.' By disavowing the idea of art making as a rarefied activity, Morris opened the door to a wealth of possible resources beyond the rigid confines of the art market. Partnerships between artists and industry were, and continue to be, one approach.

M12's collaboration with the Washington County Commissioners on a site in Last Chance, Colorado, exemplifies another strategy: partnering with other non-profit, sometimes municipal, entities. The Commissioners have leased to M12 a 40 acre parcel of grassland for a token sum. On that site, M12 has constructed its *Last Chance Module Array* (2015–16) (Figure 6.2).[6] This structure includes the fourth and fifth renditions of our Prairie Module series – the first, second, and third are located in Indianapolis, Indiana (2008–2009), and Reedsburg, Wisconsin (2012), respectively. Like *The Breaking Ring*, the *Last Chance Module Array* has a dual nature: at once formally significant, but also designed to move the viewer's attention beyond the structure – toward an event inside or to the landscape outside, toward extra-art conditions and spaces to which the work itself responds. In the enclosure created by these modules, we have hosted potlucks and star parties. At each solstice, the sun sets and rises in the center of the main module structure.

The *Last Chance Module Array* acknowledges our complex and often unex-amined relationship with the land, and the site on which the work exists is no

Figure 6.2 M12 Studio, with Onix Architects (2015–2016). *Last Chance Module Array* (Modules No. 4 and 5). Last Chance, Colorado. Photograph by Anthony Cross. Image courtesy of M12 Studio.

exception. Located one mile south of the intersection of State Highway 36 and 71, at the crossroads of Last Chance, the town received its name because it was the 'last chance' to get supplies before continuing west toward Denver or east toward Kansas. Fewer than twenty people live there. The arrangement of Prairie Module forms creates a quiet complex for experiencing the subtlety of this site. But the forms themselves are also meaningful – reminiscent of regional timber frame structures and pole barns. The timbers are blackened, having been finished with a wood-burning technique known as Shou Sugi Ban, which increases their durability. The resulting forms, like smoldering blackened frames, recall artist Robert Smithson's (1996, p.72) remarks referring to new construction along the east coast: 'This is the opposite of the "romantic ruin" because the buildings don't fall into ruin after they are built but rather rise into ruin before they are built.'

The *Last Chance Module Array* represents this dialectic too. It is a meditation on the complexities of the built landscape along the American High Plains, where the work is located, and characteristic of remote towns more generally: though populations in many rural places are surely in decline, six of the ten fastest-growing metro areas in 2013 were in the greater Plains region. The *Last Chance Module Array* is a humble intervention on the site – modest in scale compared to the vast Prairie beyond it, but also reminiscent of the structures within the vernacular landscape that surround it. Like early homestead structures, interventions like M12's *Last Chance Module Array* represent a desire for rootedness, but also reflect the present-day realities of rural lives and landscapes, which are swiftly changing, and in many cases very quickly declining. Driving along this quiet highway, one cannot miss the structures. They could be a new construction, infrastructure to support a fracking boom, or they could be the abandoned remains of a fire, which on more than one occasion has devastated the town of Last Chance.

Partnerships, like ours with the Washington County Commissioners, are characteristic of creative placemaking initiatives. Indeed, some grants require that partnerships be formalized – that project applicants or their partners have a government-issued non-profit status – a requirement that is more often restrictive for responsive practices like ours. The *Black Hornet* (2009),[7] a four-cylinder, front-wheel-drive Honda that raced at the I-76 Speedway in Fort Morgan, Colorado, in 2009–10, was a collaboration between M12 and the Hall family, who would have been ineligible for most creative placemaking grants, even though that family's efforts quite literally mobilized the value of exchanging intergenerational knowledge in order to challenge the assumption that the revitalization of a rural place depends upon the introduction of cultural expertise from somewhere else. Though M12 has benefited a great deal from collaborating with our neighbors, it would also be accurate to say that we have relied on their generosity – not least for space and borrowed machinery.

What's wrong with creative placemaking, for creatives and for places?

Artists are rural citizens too and, like rural citizens, we lack the resources required to fulfill our work: space to provide for sustained encounters between our collective and community partners for the purpose of advancing our common goals. For now, and for reasons beyond the scope of this chapter, this goal is elusive and ongoing for M12. How might we as a community of public arts professionals make it easier for artists working in this way to make *these* sorts of places. How might the field of creative placemaking as it has evolved unknowingly undermine artists' attempts to improve on their own infrastructural needs?

Collaboration in the rural case is frequently framed as 'sharing' (already limited) resources. Requirements that artists formalize their collaborations with local partners reinforces the presumption that it is acceptable to put stress on those partners. Partnering, sharing, collaborating, these approaches seem ideal in principle – it takes a village to be sure – but framing social art this way has underlying consequences. It presumes that it is up to communities, in our case frequently poor ones, to be financially solvent enough to have resources to spare and, therefore, to share. Work hard enough, work together, and good things will come. If only this were true: Only 1 per cent of arts philanthropy finds its way to rural areas in the United States, and those resources come with restrictions. It is true that demonstrated support – whether committed partners, matching grants, and so on – can validate the quality of an artistic enterprise. But it is also true that these restrictions on the carrot of creative placemaking risk shaping the work too. Working artists need less 'to-dos.' Less restrictions on their already finite resource: time. They need the time it takes to sit, to listen, to exchange ideas, and to make. They need space. Having a place for contemplative walking and image gathering, for meeting to share knowledge and stories and to workshop ideas, for eating together and for resting comfortably – these are essential for artists too, who are members of the communities in which they reside and work. The field of creative placemaking, at least in the United States, is increasingly shaping the work that is being produced because only the work for which artists are paid gets made. By paying for those works that demonstrate 'growth,' granting organizations that champion 'creative placemaking' are effectively asking artists to sing first for their dinner – a counterproductive stress on artists and their work, which are a public good.

In his 1964 book *The Machine in the Garden: Technology and the Pastoral Idea in America*, Marx (1964, p.4) wrote: 'the pastoral ideal has been incorporated in a powerful metaphor of contradiction – a way of ordering meaning and value that clarifies our situation today.' To make his point, he outlined two kinds of pastoralism: a 'pastoralism of sentiment' and a 'pastoralism of mind.' Examples of the first include leisure activities like camping and gardening, flights to the suburbs. Marx is critical: sentimental approaches to a so-called rural experience orient our attention away from the urban issues that drive one to the country. They are distractions and not solutions. The second kind of pastoralism – what he describes as a pastoralism of mind – is more complex. We find this kind of pastoralism in the arts, namely in literature as a metaphor that may 'enrich and clarify our experience' by contrasting two modes of consciousness: rural peace and simplicity with urban power and sophistication. Marx gives an example: Nathaniel Hawthorne's account of the train whistle that interrupts his quiet moment in the country. This 'little event,' also called a 'counterforce,' produces a dissonance that, once presented, demands resolution – it provides 'a check against our susceptibility to idyllic fantasies,' a reminder that sentimental pastoralism is an urban phenomenon. Today, we need not even leave the city to exercise our nostalgia for rural places – we can read *Modern Farmer*[8] while having lunch at a farm-to-table restaurant. What then are we to make of the pastoralism found among creative placemaking initiatives in a rural case? Are they sentimental or a train – no longer a metaphor for industrialization, but rather for a global cultural economy built on urban sentimentality?

Each little event, each 'creative placemaking' initiative that assumes that the quality of a rural place is built using cultural expertise from urban centers is an opportunity. But what kind of an opportunity? An opportunity to do more, without a sense for what we're doing and for whom? Without a sense for answers to these questions, creative placemaking isn't really an opportunity at all; but, rather, a pastoralism of mind that quietly siphons and redirects more resources from artists and rural communities than it returns.

Notes

1 M12 Studio, http://m12studio.org
2 *This Road Leads to Nowhere: Pierre Punk* (2016), http://m12studio.org/this-road-leads-to-nowhere.html
3 *An Equine Anthology* (2015), http://m12studio.org/an-equine-anthology.html
4 *The Breaking Ring*, http://m12studio.org/the-breaking-ring.html
5 *Gran's University*, Kultivator and M12 Studio (2012), http://m12studio.org/grans-university.html
6 *Last Chance Module Array* (2015-16), http://m12studio.org/last-chance-module-array.html
7 *Black Hornet* (2009), http://m12studio.org/the-black-hornet.html
8 Modern Farmer, https://modernfarmer.com

References

Beuys, J. (1990). 'I am searching for field character (1973)' in Kuoni, C. (ed.) *Energy Plan for the Western Man: Joseph Beuys in America, Writings by and Interviews with the Artist*. New York, NY: Four Walls Eight Windows, 21.

Burnham, J. (1968). 'Systems Esthetics', *Artforum* 7(1), September.

Lippard, L. (1980). 'Sweeping Exchanges: The Contribution of Feminism to the Art of the 1970s' in *Art Journal*, Fall/Winter, 362–365.

Lippard, L. (2016). 'Where We Are', unpublished interview with Richard Saxton in conjunction with *Where We Are*, a semester-long symposium organized by Lucy R. Lippard at The University of Wyoming.

M12 and Rothlisberger, M. (2018). *Star Route 1*. Last Chance, CO, and Rotterdam, NL: Last Chance Press and Jap Sam Books.

Markusen, A. and Gadwa, A. (2010). *Creative Placemaking*, Washington, DC: National Endowment for the Arts. Available from: www.arts.gov/ [Accessed on: December 31, 2017].

Marx, L. (1964). *The Machine in the Garden: Technology and the Pastoral Idea in America*. New York: Oxford University Press.

Morris, R. (1980). '*Notes on Art as/and Land Reclamation*', *October*, Spring 1980.

Oldenburg, C. (1961). 'I am for. . .', *Environments, Situations, Spaces*, exhibition catalogue, 25 May–23 June, Martha Jackson Gallery, New York.

Robert, S. and Flam, J. (1996). 'A tour of the monuments of Passaic, New Jersey' in *Robert Smithson: The Collected Writings*. Berkeley, CA: University of California Press.

7 Creative placemaking in peri-urban Gothenburg

Mission impossible?

Michael Landzelius and Peter Rundqvist

Abstract

In Gothenburg, Sweden, the urban development project Gothenburg Development North East (GDNE), started in 2011. The project assumed that Cultural and Creative Industries (CCIs) could serve as an instrument for the city's integration and economic development policies. The project was founded upon a mix of ideas that ranged from contradictory ambitions of the Swedish national cultural policies of 1974 and 2009, to national as well as international strategies for growth through CCIs. Although in many parts successful, the project encountered structural obstacles that hindered CCIs to play such a role in peri-urban Gothenburg, affected by steadily increasing economic, social, and ethnic segregation. After briefly situating the project in the context of a shift in Swedish cultural policy from a generalized welfare state system towards an entrepreneurial neo-liberal approach, the chapter discusses, first, the GDNE project in terms of objectives, measures, and results, and, second, relates this to ongoing discussions of creative placemaking. The conclusion draws attention to how the GDNE project illustrates the difficulty of using CCI-based alliances initiated by the public sector to solve problems in vulnerable, economically weak and socially segregated areas.

Introduction

In 2011, funding from the European Regional Development Fund (ERDF) facilitated the start of Gothenburg Development North East (GDNE, 2010). The GDNE project was a public initiative, and for the running of the project the City of Gothenburg created a new and separate publicly owned stock company. The overarching ambition was to make a difference in a quite deprived and segregated area where no private investment interests or exploitation pressures existed. With a focus on the support of entrepreneurship in general, and Culture and Creative Industries (CCIs) as well as green businesses in particular, the project objective was to promote peri-urban sustainable development in the city's north-eastern neighborhoods.

The project faced a number of challenges. The total population of this north-eastern area of the city was circa 100,000. More than 50 percent of the inhabitants have a non-Swedish background and with a strong intercultural character, this majority is a complex mix that includes more than 160 different nationalities and languages. Unemployment rates are high across age groups; social security costs are high; the educational level in general is lower than average; democratic participation is low; and many people speak little or no Swedish. With regard to young people, school results are far below average and the number of students proceeding to higher education is far below the average for the city as a whole. In the peri-urban north-east, four neighborhoods in particular were addressed in the project, all of which were mainly characterized by large apartment blocks of various design from the 1960s and 1970s.

In the following we will focus on two issues with a bearing on this case. After briefly situating the GDNE project in the context of a shift in Swedish cultural policy from a generalized welfare state system towards an entrepreneurial neo-liberal approach, the chapter discusses, first, the project in terms of objectives, measures, and results, and, second, relates this to ongoing discussions of creative placemaking.

Cultural policy meets socio-economic restructuring

The GDNE was born in a cultural climate that in 2009 finally led to a legislative shift in terms of Swedish cultural policy. Up to this point, cultural policy was informed by eight objectives passed unanimously by parliament in 1974 (Swedish Cabinet, 1974, p.292–98). The policy objectives concerned reaching all citizens, geographically and socially, as evenly as possible; artistic freedom and independence; opportunities for personal creative activity; protection of culture heritage with a requested expansion of heritage to also represent the lives of ordinary people; and the promotion of 'exchange of experiences and ideas in the field of culture across linguistic and national boundaries.' However, in hindsight and looking back through the lens of neo-liberal marketization, one objective clearly distinguished the 1974 cultural policy from others – the goal to 'counter the negative effects of commercialization in the field of culture.' That this objective was unanimously supported by parliament indicates the extent to which commercial culture at the time was seen as insufficient with regard to the creation of cultural expressions with authenticity or value.

However, while political majorities continued to shift over a period of more than thirty years in which Sweden saw many market-oriented reforms and an extensive rolling back of the welfare state, much remained the same in the area of culture until 2006. This year a review of Swedish cultural policy was initiated by the governing center-right coalition that resulted in a neoliberal shift, including in cultural policy. To 'counter negative effects of commercialization' was now replaced with the wording that '[t]here is no inherent contradiction between economic viability and artistic quality or freedom' (Swedish Cabinet, 2009, p.28). In this spirit, a separate section on Cultural and Creative Industries and Regional Growth (ibid., p.72) appeared together with a recurring stress on entrepreneurship throughout. This break in Swedish cultural politics was explicitly noted in a parliamentary motion made by the Social Democrats:

> The present cultural policy objectives were passed unanimously thirteen years ago, in 1996. The commission and official report that preceded the vote included all parties in parliament, just as was the case with the passing of the cultural policy of 1974. It is indeed regrettable ... that the bourgeois coalition cabinet this time chose not to have a parliamentary commission and instead chose to break with this tradition. Accordingly we move for a rejection of the proposed new cultural policy objectives
>
> (Pagrotsky et al., 2009, p.4).

This cultural policy context and clear historical shift is where the question of the role of culture in disadvantaged suburbs is situated. In the 1960s and 1970s, when the large suburban areas were built, cultural policy was based on the model of the state as architect and social engineer (Frenander, 2001; Hillman-Chartrand and McCaughey, 1989). This policy sought to provide all citizens with equal access to culture and cultural activities and projects were publicly funded. The building of suburbs such as the ones addressed in the GDNE project – with spaces for suburban leisure-time ranging from sports facilities to public libraries and cultural centers – exemplifies a form of creative welfare state placemaking that was enabled through a combined 'spatial fix' (Harvey, 1982, pp.429–38) and 'institutional fix' (Uitermark, 2005, p.159) resulting from a long-term agreement between labor and capital (Larsson et al., 2012a, 2012b; Peterson, 2012). The period between 1974 and 2009 saw Sweden entering into neo-liberal deregulation (Blyth, 2001; Larsson et al., 2012a) and a shift towards 'governing without Government' (Rhodes, 1996, pp.660–67), or 'governance-beyond-the-state' (Swyngedouw, 2005, pp.1993–95), involving a 'quango-ization' of the state (Miller and Rose, 2008, p.213). With the 2009 cultural policy, deregulation finally came to reframe culture, affected by international turns towards the 'creative economy' and notions to replace industrial production with knowledge-based industries and services (Doel and Hubbard, 2002; Duxbury et al., 2016; Florida, 2002; Hutton, 2008; UNESCO, 2013). The basis for this turn towards the 'creative' was essentially that '[i]n the production of all kinds of goods and services, the symbolic dimension has gained central importance, which has led to an almost universal aestheticization of goods and everyday practices thanks in particular to publicity and design' (d'Ovidio and Rodríguez Morato, 2017, p.4; Lash and Urry, 1994). While policy prescriptions, made by proponents of this 'creative turn,' have focused on the potential of cultural practices as socio-economically and politically integrative and equality-building in cities, regions, and countries, much critique has addressed the fact that practical implementations show questionable results (Grodach, 2017; Lindeborg and Lindkvist, 2013; O'Connor, 2016).

Importantly, it has been pointed out by Grodach et al. (2017) that these trends need to be seen, not as neutral, but as interest-based forms of harnessing and channeling urban potentialities that leave out progressive socio-spatial options of linking artistic and knowledge production with manufacturing. Such criticism is perhaps particularly relevant in a city such as Gothenburg, with its long industrial history. They also note that presently dominating strategies lead to zoning regulations that enforce the removal of any industrial activities in favor of clean, so-called creative industries. Essential to note here is that whichever kind of place-making we choose or confront, it carries within itself an unavoidable mirror image: the creative destruction of earlier social and built forms, as well as the exclusion of other options: 'Creative destruction is embedded within the circulation of capital itself,' and 'masses of capital and workers shift from one line of production to another, leaving whole sectors devastated, while the perpetual flux in consumer wants, tastes, and needs becomes a permanent locus of uncertainty and struggle' (Harvey, 1989, p.106). This shows the instability, insecurity, and constant search and creation of new markets, including real estate markets, for newly invented products and uses that characterize creative destruction. The rolling-out of the 1960s and 1970s areas we see in north-eastern Gothenburg and the simultaneous expansion of the social and cultural programs of the welfare state came when Sweden reached its zenith as an industrial-based economy. Just like the present rolling back and revamping of those solutions, the process was double-sided and included a mirror image of socio-spatial destruction. While the notion comes from Marx, the Austrian-American political economist Joseph Schumpeter was equally at home with the idea that creative destruction is an inherent part of capitalism (Harvey, 2011, p.46; Greenspan, 2005). The point we wish to make is that there is nothing innocent nor anything necessarily 'creative' or 'progressive' about the 'creative economy' seen in a wider context of urban policy and urban justice.

Gothenburg Development North East

Across the political spectrum, the Gothenburg City Council recognized that economic and ethnic segregation, increasing social inequalities, and high unemployment, were serious problems in Gothenburg, and they continue to be so. In this political context, the Gothenburg Development North East project was initiated. The GDNE project was thus a public initiative. For the running of the project the City of Gothenburg registered a new and separate publicly owned stock company – Utveckling Nordost AB (Development North East Co). The board of this company was politically appointed and consisted of members from the City Council. The project partners consisted of seven different departments within the city administration plus the city's business development office, Business Region Göteborg (another non-profit stock company owned by the city). The international auditing and consulting firm PricewaterhouseCoopers was selected as official project evaluator (PwC, 2013). It is obvious that this set-up fitted well with the expectations and priorities of the new cultural policy of 2009. Nonetheless, there are political tensions in Gothenburg, where a City Council majority coalition formed by the Social Democrats, the Greens, and the Left Party has been in place since the fall of 2010. In terms of cultural policy, this majority does not represent a neo-liberal turn but rather holds on to many of the values of the 1974 policy.

Objective and the role of culture

In the design stage, the GDNE project faced the facts outlined above, such as high levels of unemployment, rising social security costs, falling school results, and low levels of democratic participation. The overall objective of the GDNE project was to contribute to sustainable urban development in Gothenburg with a focus on the peri-urban north-eastern neighborhoods (GDNE, 2010). The project was planned to run from 2011 to 2013, and upon acceptance from the ERDF received a budget of 123 million Swedish krona (SEK) (12.8 million euros) of which the ERDF contributed 40 percent and the City of Gothenburg the remaining 60 percent. Identified goals were to develop the business sector, create jobs, develop cultural life and the urban environment, and to increase the attractiveness of the north-eastern parts of the city. The primary target group consisted of three partly overlapping categories: the inhabitants, those working in the area as well as people conducting business there. The project's activities were divided into four themes: Business, Culture, Urban Environment, and Vision and Communication. The themes were mutually interlinked and the efforts were planned to reinforce one another: Business aimed to promote business development in general; Culture focused on gender equality and on the promotion of entrepreneurial approaches within the arts. The Urban Environment theme focused on accessibility and mobility issues, including increased movement between places of shared utility and residential areas, and on personal safety and security. The Vision and Communication theme aimed to increase the attractiveness of north-eastern Gothenburg. Of the 123 million SEK budget, Business was allocated 10 percent, Culture 26 percent, Urban Environment 37 percent, and Vision and Communication 11 percent (actual turn-out stayed close to predicted). The remaining funds went to project management. It should be noted that while the project supported entrepreneurship in general, the particular focus was on CCIs and green businesses. The role of culture was thus central to the overarching ambition.

When the application was drafted, there was an assumption that 'culture' would contribute to cohesion, integration, business development, and jobs. There was a double ambition, both to break new ground in terms of attitudes and relationships between suburban residents and to translate 'culture' into business and jobs. The project design thus reflected general developments in the European Union (European Commission, 2012) and elsewhere (UNESCO, 2013). While CCIs make up only a small part of the economy, there were also great expectations within the GDNE project that CCIs would become engines for post-industrial economic development. A pronounced CCI-policy is yet to be developed in Gothenburg, and CCI investments and development projects have been concentrated to close-to-the-center areas, e.g. activities related to the publishing and film industry have been provided with facilities in the revamped port and shipyard areas that now form part of the center and are exposed to high exploitation pressures. The ambition of the GDNE project-design to establish peri-urban CCIs in a context of socio-economic and ethnic segregation was thus an innovation. The project also noted that CCIs in peripheral locations could contribute to local suburban gentrification by having a negative impact on the life and positive dynamics of existing communities and their diversity (Figure 7.1).

Figure 7.1 Apartment building in the Hammarkullen neighborhood with a windowless concrete gable remade into a public work of art: *Every time I look up I feel happy, when I look down I'm sad*, by the British artist Ben Eine. This particular piece was made in 2016 during the Gothenburg Street Art Festival Artscape, after the initial ERDF-supported period of the GDNE project was over. The building is owned by the city's housing company Bostadsbolaget. Photo: Michael Landzelius.

Obstacles, incubators, startups

Four neighborhoods in north-eastern Gothenburg were involved in the GDNE project: Hammarkullen, Angered, Gårdsten, and Rannebergen. They are located 12–16 km from the city center, and travelling times by tram and bus from the city center to central stops in these areas vary between 17–35 minutes. While these are quite fast connections compared to the situation in many cities, the distance is not only physical but also social as well as mental – experienced as a rupture in the urban fabric between the deprived peri-urban neighborhoods and the creative zones of the city center, and representing a hinderance to the flow of people and the exchange of ideas, as well as to the will to invest capital. In the performance of the GDNE project, other obstacles were also identified that challenged the engagement of local inhabitants as well as potential entrepreneurs and the role of culture as a cohesive and business-creating factor. For example, the bureaucratic and slow administration of individual unemployment cases and of project applications created frustration and made potential cultural entrepreneurs hesitant to try to realize their ideas. In an attempt to alleviate or even remove some of these structural constraints, the GDNE project launched a series of culture-incubators in order to initiate, develop, and support cultural entrepreneurship. One immediate result of these efforts was the startup of about twenty new small CCI companies, the majority of them run by immigrants (PwC, 2013). However, on a more general level, while unemployment did go down during the performance of the project, in the years directly after the GDNE project Sweden and Gothenburg underwent an economic boom that may be considered a more important factor for reduced unemployment rates, increased investments, and increased social stability in the north-eastern neighborhoods. In the Angered district unemployment rates fell from 10.9 percent in 2009 to 9.7 in 2016, and in Hammarkullen from 16.4 percent in 2009 to 14.0 in 2016.

Urban art network

One of the activities supported by GDNE was the Urban Art Network, a sub-project aimed at creating a multidisciplinary network for both arts education and existing cultural activities in north-eastern Gothenburg. The purpose of the network was to enable the arts to contribute to the segregated city's needs and opportunities through specific targeted development projects, pilot projects, and the creation of new learning environments. Based on the educational and research environment of the district, the network was to support and contribute to artistic practices, for example community art and cultural heritage work. The Urban Art Network focused on three main targets: firstly, broader recruitment to higher arts education through better collaboration between arts education programs; secondly, arts development; and thirdly, sustainable community development on local, urban, and regional levels. To reach the first core target, cooperation had to be established between the city administration, the regional administration, art schools, and the University of Gothenburg so that preparatory courses and admission criteria were harmonized, and also in order that some educational institutions and programs at higher and research levels could be located outside the university's city center facilities and anchored in the north-eastern areas of the city. Activities resulting from the GDNE project were still alive and expanding in 2017, such as education for music teachers (*The Music College*), an MFA program in Literary Composition, and a one-year preparatory course in Performing Arts, all of them organized in collaboration with the Faculty of Fine Arts at the university. An average of six students each year have graduated from the course for music teachers, and they have either found work as music teachers or proceeded to higher music studies.

Investments in places for culture

An important undertaking made as part of the GDNE project was the renovation of a local cultural center with libraries, a theater, a movie theater, an art gallery, meeting rooms, and a restaurant. The center was built in 1979 and was the first of its kind in a Swedish suburb. The renovation was funded by public money and aimed to make the center more attractive to more groups of citizens. The project had no significant effects on the CCI sector, and no effects on the labor market for artists have been discussed or registered, but the main objective to make the center more attractive was reached, with a remarkable rise in the number of visitors.

The GDNE project also supported several smaller initiatives for creating new meeting places and activity centers, the intention of which was to support local community life and enable more citizen initiatives. Such local placemaking initiatives have especially occurred in the district of Hammarkullen, where unemployment rates remain constantly high, but where the civil society sector is lively and mobilizes inhabitants in different cultural development activities, from local arts projects and a yearly carnival to political engagement. The many examples also of minor sub-projects among the GDNE initiatives have strongly contributed to an increase in the level of trust among citizens with regard to local cultural institutions and in the value of activities supported by the public sector. This may to some extent have contributed to a rising level of interest, particularly among young people in the area, in engaging in civil society and social and cultural activities, many of them related to the fighting of segregation.

PricewaterhouseCoopers' review in brief

The overall objective of the project was, as we have described above, to develop north-eastern Gothenburg with a focus on four themes – Business, Culture, Urban Environment, and Vision and Communication. In their review of the project, PricewaterhouseCoopers (PwC, 2013) noted that within these four themes a total of eighteen different overarching activities were carried out, ranging from investments in physical interventions in the environment to development work for and by the local business community. Measurable objectives set by PwC at the outset of the GDNE project included the following goals that were to be achieved by the end of the project: 220 new jobs (with equal gender distribution) located in north-eastern Gothenburg; sixteen new companies (again with equal gender distribution) in the area; five investments for increased attractiveness of the area; and eighty-five companies participating in the various efforts within the project. To measure these four objectives, 343 activities with specified goals were defined– of these, 338 were reached. One of the five goals that was not reached was the number of created jobs, but the analysis noted difficulties with directly linking some jobs that were created to the performance of the project as such. The actual results of the four main objectives were: 123 jobs; fifty-six start-ups; five attractiveness-investments; and 258 participating businesses. From the point of view of PwC as project evaluator, the GDNE project was indeed a success. The City Council decided in 2014 that the project in revamped form should continue to work on urban development strategies as well as attempt to seek further EU funding, now under the management of Business Region Göteborg.

The GDNE project – creative placemaking?

The GDNE project had a larger budget than any other comparable project in Sweden has ever had. The project resulted in a mix of activities that all had the ambition to contribute to dynamic creative placemaking, strengthening civil society, and reducing segregation locally. One characteristic common to all these initiatives was that they originated 'from outside and above.' The initiatives came from public sector bodies with good ambitions. In accordance with the 'multiple-helix' development models already forecasted in most European policy tools and long-term strategies for lifelong learning, for economic and social cohesion, and for sustainable development, the project mobilized existing collaboration – between different public administrative bodies, between educational institutions including the university level, and between different jurisdictions (from the City of Gothenburg and adjacent municipalities to the region). The GDNE project stressed capacity building with regard to the shaping of new strategies to counter segregation, to create jobs and to build cross-cultural ties. The peri-urban north-east is a vivid example of the global as both eminently glocal and complex: it is an area characterized by people of more than 150 nationalities and with multiple kinds of challenges: demographic and generational shifts; economic inequalities; segregation; extensive in-migration; complex conflicts structured by ethnic, religious, and class differences, and with some groups and individuals that represent beliefs highly negative to boundary-crossing and hybridity (World Values Survey, 2016; Welzel, 2013).

Given this, an ambitious creative placemaking project of the kind undertaken by GDNE will have to work with an innumerable set of unknowns in terms of reception among target groups, in this case first and foremost the inhabitants of 150 nationalities, and those working as well as conducting business in the area, many of which were also inhabitants. As we have noted, the project emerged out of public sector bodies that intended the project to be democratic and emancipatory. Yet, partly because of the kinds of interventions such an agenda may lead to – particularly when culture is mobilized as a tool at the same time as culture is a contested issue in the complex local community – well-intended projects might backfire and lead to local failures and increased distrust between ethnic and religious groups and individuals as well as between local citizens and project initiators. Indeed, the project was often met with skepticism by citizens who felt they did not have much influence over the plans and activities, which they believed did not build upon their real needs and wishes. The extent to which favoritism and intercultural conflict and other such issues were experienced is yet to be researched.

In a much-quoted text, Markusen and Gadwa (2010, p.3) have defined creative placemaking as a situation when 'partners from public, private, non-profit, and community sectors strategically shape the physical and social character of a neighborhood, town, city, or region around arts and cultural activities.' This definition fits with the GDNE project, with its stress on the Culture theme and the fact that CCIs were planned to play a major importance also in the performance of activities in the other three themes. However, we wish to make a point here based on a slightly longer quote from Markusen and Gadwa (ibid., pp.5–6) on the core 'six components of a successful strategy.' Such a strategy, they write:

> starts with an entrepreneurial initiator; demonstrates a commitment to place and its distinctive character; mobilizes public will, both in local government and the citizenry; attracts private sector support, either from cultural industries or place developers or both; wins the active participation of arts and cultural leaders; and succeeds in building partnerships across sectors (for-profit, non-profit, government, and community), missions (e.g., cultural affairs, economic and workforce development, transportation, housing, planning, environment, and health), and levels of government (local, state, and federal).
>
> (Markusen and Gadwa, ibid., pp.5–6)

The point we wish to make is that the GDNE project failed from the outset with the first component in that it did not start with an entrepreneurial initiator but with local government (i.e. half of the third component); it succeeded with the second and the fifth; partly also with the last; but failed with the fourth and the last since support from major private sector interests did not materialize. In this, the project partly resembles an 'older' form of cultural policy practices based on local state initiatives and control. However, we would argue, this resemblance was not desired by the City Council majority, nor by project management or the local state neighborhood bodies, but resulted because the suburbs in north-eastern Gothenburg, with their particular needs, problems, and geographical locations, have poor competitive characteristics in relation to other parts of the cityscape (Landzelius, 2012). We can thus note that the first component of a successful placemaking strategy in the view of Markusen and Gadwa was not possible to fulfil in Gothenburg's peri-urban north-east. One might argue that this is a too literal reading of Markusen and Gadwa, and we will return to that shortly. Nonetheless, the case shows how a City Council majority can take a council-decision to start a project such as the GDNE, but then what? With the rolling back of the welfare state and the lack of entrepreneurial interest, what remains to be done?

Before we conclude, another piece in the puzzle is worth mentioning, which could open up a discussion of ripple effects that might be related to a project such as the GDNE. This piece follows upon the cabinet's initiative in the fall of 2016 to create a special governmental body, effective from January 1, 2018, for a coordinated and efficient implementation of the national policies on gender equality (Swedish Cabinet, 2016). After parliament's passing of the bill in December 2016, the cabinet decided that this new Swedish Gender Equality Agency was to be located in Gothenburg, and half a year later that the agency was to be placed in Angered. The Minister for Gender Equality, Åsa Regnér, noted: 'That the agency is placed in Angered and not centrally in Gothenburg, or in Stockholm, is in accordance with the cabinet's objective that state jobs and work places should be found in different parts of cities and across the country'; while the Minister for Social Affairs, Ardalan Shekarabi, commented on the same topic: 'Neither people, places, suburbs, or entire areas of the country, are to feel abandoned' (Swedish Cabinet, 2017). The Swedish Gender Equality Agency will from the start have seventy-five employees. The long-term effects with regard to ambitions such as the ones governing the GDNE project cannot be predicted, and comments from politicians, academics, and the press vary with regard to its impact on jobs as well as the local image.

Conclusion

The final report from PwC noted that the achievements of the project needed to be taken care of, nurtured, and continued to be built upon. Some such results and areas of activity clearly belonged to one of the seven participating administrative bodies of the city and could be fused with their normal responsibilities, while other results ended up – as in many project-based temporary activities and interventions – falling between the cracks. There are a number of remaining questions facing the GDNE project: What kinds of jobs and education were created in terms of local relevance and long-term sustainability? What kind of long-term sustainable impact was envisioned and planned at the outset, and what was achieved and could at the end be secured through the project format? What kinds of local expectations did the project explicitly as well as implicitly raise, and how should discontent and failed expectations be met when the project has been dismantled? How did the project's belief in CCIs fit in relation to an area such as north-eastern Gothenburg, in terms of both the needs of its extremely diverse population and the competitive characteristics of the area in relation to other parts of the cityscape?

We would argue that the GDNE project illustrates the difficulty of using CCI-based alliances initiated by the public sector to solve problems in vulnerable, economically weak, and socially segregated areas. While this was a huge under-taking with a budget of 123 million SEK, public funding – be it from the local state, the central state, or the EU – can only reach so far. The project could be seen as a mix of both the cultural policy of 1974, in which commercial culture was seen as insufficient and which thus aimed to 'counter the negative effects of commercialization in the field of culture,' and the policy of 2009, the latter supported by the fact that on-the-ground conditions today are significantly more market-oriented. This mix in the GDNE project could be seen as a way of integrating important values not present within neo-liberal agendas. With regard to the quite literal reading of Markusen and Gadwa above, we might perhaps indeed argue that the local authority – with additional economic support from the ERDF – effectively and with good results acted as entrepreneurial initiator. While this is true in a sense, the results of the GDNE project show that in a society where everything is expected to run on market conditions after the end of a project phase, that phase needs to secure strong private entrepreneurial interests if the accomplishments of public investments are to survive. The remaining bottom-line question thus concerns what can be done when the public sector has to gradually pull back.

The conclusion we have reached based on this study is that neoliberal-influenced confidence in CCIs may be relevant in areas that at some stage finally are recognized as having economic potential in the eyes of entrepreneurs – for some, the city is, after all, not much more than a growth machine (Molotch, 1976). In this period of 'quango-ized' governance, creative placemaking seems to work in spaces and situations where cultural practices can be made to overlap with CCIs and investment interests tied to local and regional agendas of economic growth and entrepreneurship. This is the case in Gothenburg's more central parts, as in so many other cities around the world. In such situations, creative placemaking may well, in very real terms lead to both creative (de) construction, with all its accompanying spatial and social consequences in terms of its often-observed contribution to gentrification and segregation. In poor and weak areas such as north-eastern Gothenburg, the idea of creative placemaking is at risk of remaining exactly that, an idea or a form of 'therapeutic' activity that promotes false consciousness and ideological *méconnaissance* in a perhaps both Marxian and Lacanian sense of concealing both social contradictions and the subject's self-recognition (Harvey, 1989; Lacan, 1977). Creative placemaking, when seen as a processual whole, is in this sense something that is always struggled over, open to various forms of articulation in which moments and parts of the whole are controlled in unequal ways by different social actors.

This brief analysis of the GDNE project has clearly shown that publicly initiated creative placemaking of this kind is not enough to create sustainable structural changes in vulnerable areas. We believe that more in-depth research is needed in order to understand the possible role of culture and the arts as well as CCIs as contributors to sustainable urban development in the context of suburban and peri-urban neighborhoods such as the ones in north-eastern Gothenburg. Research on these issues needs to engage both with a critical scrutiny of governmental policies and practices ranging from cultural policies to securitization, and with asking questions about if and how present forms of quango-ized governance could be harnessed for channeling entrepreneurial interests and investment decisions to areas that seemingly lack competitive advantages. This in a situation where inter-urban competition seems to have turned more destructively intra-urban than ever before. In this regard, the GDNE project was an ambitious attempt to seek a path away from such intra-urban competition, yet simultaneously sought to make Gothenburg more competitive on the inter-urban stage. On the one hand, research needs to focus on the realism and ideological conundrums of such strategies, and on the other it should not forget to address how groups and individuals from different cultural zones understand, embody, and relate to identity, social change, and democracy in relation to the interventions of CCIs and cultural artistic practices, in order to make innovation in peri-urban neighborhoods feed long-term social sustainability and peace rather than run the risk of feeding tensions and conflict.

References

Blyth, M. (2001). 'The transformation of the Swedish Model: Economic ideas, distributional conflict, and institutional change', *World Politics*, 54.

d'Ovidio, M. and Rodríguez Morato, A. (2017). 'Introduction to Special Issue: Against the creative city: Activism in the creative city: When cultural workers fight against creative city policy', *City, Culture and Society*, 8.

Doel, M. and Hubbard, P. (2002). 'Taking world cities literally: Marketing the city in a global space of flows', *City*, 6.

Duxbury, N., Hosagrahar, J., and Pascual, J. (2016). 'Why must culture be at the heart of sustainable urban development?' (Agenda 21 for culture – Committee on Culture of United Cities and Local Governments). Available from: www.agenda21culture.net/sites/default/files/files/documents/en/culture_sd_cities_web.pdf. [Accessed: 6 September 2017].

European Commission. (2012). *Promoting cultural and creative sectors for growth and jobs in the EU*. Communication from the Commission to the European parliament, the Council, the European Economic and Social Committee and the Committee of the Regions. /* COM/2012/0537 final */

Florida, R. L. (2002). *The Rise of the Creative Class: And How It's Transforming Work, Leisure, Community and Everyday Life*. New York, NY: Basic Books.

Frenander, A. (2001). 'Svensk kulturpolitik under 1900-talet', *TijdSchrift Voor Skandinavistiek*, 22.

GDNE. (2010). *Gothenburg Development North East* [project application to the European Regional Development Fund], City of Gothenburg.

Greenspan, A. (2005). 'Greenspan on Schumpeter's "creative destruction"', extract from: *Testimony before the Senate Banking Committee*, July 21, 2005. Available from: https://www.youtube.com/watch?v=CVALfc-nayY. [Accessed: 6 September 2017].

Grodach, C. (2017). 'Urban cultural policy and creative city making', *Cities*, 68.

Grodach, C., O'Connor, J., and Gibson, C. (2017). 'Manufacturing and cultural production: Towards a progressive policy agenda for the cultural economy', *City, Culture and Society*, 10.

Harvey, D. (1982). *The Limits to Capital.* Oxford: Basil Blackwell.
Harvey, D. (1989). *The Condition of Postmodernity.* Oxford: Blackwell.
Harvey, D. (2011). *The Enigma of Capital and the Crisis of Capitalism.* Oxford: Oxford University Press.
Hillman-Chartrand, H. and McCaughey, C. (1989). 'The arm's length principle and the arts', in Cummings, M. C. and Davidson-Schuster, J. M. D. (eds), *Who's to Pay for the Arts?* New York: American Council of the Arts.
Hutton, T. A. (2008). *The New Economy of the Inner City: Restructuring, Regeneration and Dislocation in the Twenty-First-Century Metropolis.* New York: Routledge.
Lacan, J. (1977). 'The mirror stage as formative of the function of the I as revealed in psychoanalytic experience', in Lacan, J., *Écrits: A Selection.* London: Tavistock.
Landzelius, M. (2012). 'Real estate ownership concentration and urban governance', in Larsson, B., Letell, M., and Thörn, H. (eds), *Transformations of the Swedish Welfare State.* Basingstoke: Palgrave Macmillan.
Larsson, B., Letell, M., and Thörn, H. (eds) (2012a), *Transformations of the Swedish Welfare State,* Basingstoke: Palgrave Macmillan
Larsson, B., Letell, M., and Thörn, H. (2012b). 'Transformations of the Swedish welfare state: Social engineering, governance and governmentality', in Larsson, B., Letell, M., and Thörn, H. (eds), *Transformations of the Swedish Welfare State.* Basingstoke: Palgrave Macmillan.
Lash, S. and Urry, J. (1994). *Economies of Signs and Space.* London: Sage Publications.
Lindeborg, L. and Lindkvist, L. (2013). *The Value of Arts and Culture for Regional Development: A Scandinavian Perspective.* London: Routledge.
Miller, P. and Rose, N. (2008). *Governing the Present: Administering Economic, Social and Personal Life.* Cambridge: Polity.
Molotch, H. (1976). 'The city as a growth machine', *The American Journal of Sociology,* 82.
O'Connor, J. (2016). 'After the creative industries: cultural policy in crisis', *Law, Justice & Global Development,* 1.
Pagrotsky et al. (2009). 'Motion till riksdagen 2009/10: Kr1 av Leif Pagrotsky m.fl. (s)', the Swedish Parliament, Stockholm.
Peterson, M. (2012). 'Pathways of the welfare state: Growth and democracy', in Larsson, B., Letell, M., and Thörn, H. (eds), *Transformations of the Swedish Welfare State.* Basingstoke: Palgrave Macmillan.
PwC. (2013). *Utveckling Nordost – Sveriges största stadsutvecklingsprojekt* [project evaluation report to the European Regional Development Fund], Tillväxtverket, Stockholm. Available from: www.utvecklingnordost.se/Bilder/Slutrapport%20pwc.pdf [Accessed: 23 October 3 2017].
Rhodes, R. (1996). 'The new governance: Governing without government', *Political Studies,* XLIV.
Swedish Cabinet. (1974). *Proposition 1974:28: Kungl. Maj:ts proposition Angående Den Statliga Kulturpolitiken,* Stockholm.
Swedish Cabinet. (1996). *Regeringens proposition 1996/97:3: Kulturpolitik,* Stockholm.
Swedish Cabinet. (2009). *Regeringens proposition 2009/10:3: Tid för kultur,* Stockholm.
Swedish Cabinet. (2016). 'Jämställdhetsmyndigheten placeras i Göteborg' (press release December 15, 2016). Available from: www.regeringen.se/pressmeddelanden/2016/12/jamstalldhetsmyndigheten-placeras-i-goteborg/ [Accessed: 2 January 2018].
Swedish Cabinet. (2017). 'Regeringen positiv till myndighet i Angered' (press release May 15, 2017). Available from: www.regeringen.se/pressmeddelanden/2017/05/regeringen-positiv-till-myndighet-i-angered/ [Accessed: 2 January 2018].

Swyngedouw, E. (2005). 'Governance innovation and the citizen: the Janus Face of governance beyond-the-state', *Urban Studies*, 42.

Uitermark, J. (2005). 'The genesis and evolution of urban policy: A confrontation of regulationist and governmentality approaches', *Political Geography*, 24.

UNESCO. (2013). *Creative Economy Report 2013*, United Nations, New York.

Welzel, C. (2013). *Freedom Rising: Human Empowerment and the Quest for Emancipation.* Cambridge: Cambridge University Press.

World Values Survey. (2016). 'Findings and insights: Inglehart-Welzel cultural map'. Available from: www.worldvaluessurvey.org/WVSContents.jsp?CMSID=Findings [Accessed: 25 March 2016].

8 A conversation between a collaborating artist and curator

Placemaking, socially engaged art, and deep investment in people

Jim Walker and Shauta Marsh

Abstract

After fifteen years of learning and building on its experiences, Big Car Collaborative—a socially engaged art and placemaking organization and studio led by artist Jim Walker and curator Shauta Marsh—is delving deeply into what it means to be an integral part of a community in the urban core of Indianapolis, Indiana, USA. This conversation between Walker and Marsh—which took place during a driving trip to multiple North American industrial 'rust belt' cities—explores the ideas of placemaking and social practice art and how these work together. The conversation also examines the path of Big Car's learning by experience with a variety of projects and programs to reach a better—but constantly evolving—understanding of how deep investment of time and care for a community and its people can support a stronger society. Further, they look at the roles of socially engaged art and active, program-based, and site-specific placemaking in these efforts.

Introduction

After more than a decade of learning and building on its experiences, Big Car Collaborative[1] – a socially engaged art and placemaking organization and studio led by artist Jim Walker and curator Shauta Marsh – is delving deeply into what it means to be an integral part of a community development in the urban core of Indianapolis, Indiana, USA. These collaborators, along with other artists in the organization and a variety of civic and cultural partners, are working to accomplish the goal of supporting a more vibrant community without creating a situation that prices out existing residents or the artists now moving in to fill vacant homes in the neighborhood. This work comes after learning by experience that gentrification is best handled by maintaining control of real estate and being at the table – even serving as a partner – during redevelopment efforts.

Big Car Collaborative began as a small, cross-genre collective of artists interested in surrealism and experimentation in the then-struggling Fountain Square neighborhood[2] near the city's center. Four years into operating an all-volunteer grassroots arts organization, Walker and Marsh found that they weren't satisfied with focusing mainly on supporting artists and serving an arts audience. They weren't satisfied with art as an end in itself. Big Car began, instead, to see art as a means for benefiting individuals by bringing the spark of creativity to their lives and supporting stronger communities by boosting positive perception and vibrancy through an eclectic array of programs that engage audiences in arts-focused social experiences. A turning point came in 2008 when Big Car began handing its studio and gallery over to visitors, showing collages and drawings they made in equal status with work made by self-identified artists. Big Car's lead artists saw, from this, that the audience of collaborators became better connected with Big Car and with art.

Big Car artists also began taking further steps toward working with diverse community members as collaborators – choosing to take these approaches to co-creation from institutional settings (galleries and museums where art is expected) to places where art is a surprise and where opportunities to participate in creative projects are rare. The organization progressed to libraries, schools, and community centers and then to parks, fairs, apartment complexes, and the streets – building out a mobile art unit and devising partnerships (for example, a library bookmobile) to make this click.

This work led to stronger relationships with the community and better acceptance of Big Car's ideas. This approach of collaborating with and working for partner communities is a critical element to all of Big Car's placemaking projects – from putting on arts programming in parking lots in Fountain Square on the artists' own dime to taking over Monument Circle, the most visible place in the city, for a $400,000 National Endowment for the Arts-funded[3] partnership with the City of Indianapolis.

This conversation between Walker and Marsh – who are also married and parent two teenagers – explores the path of learning through a variety of projects and programs to an understanding that a deep investment of time and care in the people of a community is the best way to truly support thriving neighborhoods through socially engaged art and engaging, programming-based creative place-making. Marsh and Walker discussed these things during an extended family driving trip to explore art spaces and public places in Pittsburgh, Buffalo, Niagara Falls, Toronto, small-town Canada, Detroit, and Fort Wayne, Indiana – before returning to Indianapolis.

The conversation

MARSH: Let's start out by talking about what we think placemaking is.

WALKER: We were just staying at a hotel near the northern shore of Lake Eerie in Canada. A newer place, it was furnished with ping pong tables, a badminton court, and other games in an indoor atrium adjacent to the swimming pool. This is sort of an indoor court outside the interior-facing rooms. The floor out there is covered in artificial turf. There are tables and chairs scattered about. People working in placemaking like we do would probably not consider that a placemaking project. It is really just a game room or activity space, the kind of thing that has been around a long time. But, when those things – ping pong, turf, chairs – are placed outside in a public space, these are the regular trappings of placemaking zones.

MARSH: Similarly, we were just talking about how we've noticed artists calling what they're doing placemaking because they've moved activities they'd normally do inside, outdoors to public or semi-public space. So instead of painting a portrait with a model they hired in their studio they do that same thing out on a sidewalk. Then they might share a picture of this on Instagram and hashtag it #CreativePlacemaking. But creative placemaking is much more than the presence of an artist making or displaying art in public. Doing art in public view is what has happened forever with *plein air* painting and at art fairs. And simply putting art out in public places for the people to experience and passively enjoy has happened in all kinds of traditional public art, also for a long, long time. Those things are great and important roles for artists, but they aren't what I see as creative placemaking.

WALKER: So, following this logic, is it accurate to simply put games outside and call that placemaking? Again, most of us wouldn't call the setup inside the hotel where we stayed placemaking. It's just the game area. But there's a difference. And this comes when the games and turf and chairs and tables serve purposes for people in public places. Placemaking really happens when people utilize recreational, social play and lingering spaces – and artists doing art – in ways that actually help solve a design problem. That's kind of what the hotel atrium games are for too. You have families stuck there for the night. They want some things to do where they are staying – without getting back in the car after a long day of driving. So they hang out in the atrium, swim in the pool, and play some ping pong. If the hotel were located, instead, in the little town by the park or closer to the lakeshore like hotels used to be before we drove everywhere, then people wouldn't need to hang out inside the hotel. They'd walk out and do stuff at other places nearby. This hotel is located in a place easy to drive to but terrible for walking anywhere. So it's a design problem. If the developers had been able to put the hotel in the right place by the lake, the game area and even the pool might be unnecessary. But this hotel – like so many things built today – is isolated. So they had to make it a self-contained, inwardly oriented place, an island. This is much like what we see with indoor shopping malls in suburbanized North America. If everything is offered in a one-stop shop, you don't need to wander by foot or bike from place to place and you don't experience spontaneous social encounters, you don't run in to your neighbors or meet new people, you don't make new discoveries. Your life becomes routine.

MARSH: I see placemaking as an effort to undo these terrible physical and social design choices. Creative placemaking is an effort by artists to be part of this work of place fixing, and also part of undoing the social damage that these choices have caused for communities.

WALKER: Right. For this reason, all actual, effective placemaking is closely aligned with the ideas of *Tactical Urbanism* – described by Mike Lyon and

Tony Garcia (2015) as 'short-term actions for long-term change'. The tactical aspects of placemaking are vital. Artists simply making art outside are not doing creative placemaking unless they are utilizing this art experience as an avenue for social reconnection or to address other challenges in direct response to and collaboration with the community where they are working. Moving games outside to help bring people together does a lot more than just moving an artist outside to paint in full view of us all like a sort of artist zoo. At least we're all invited to play ping pong when the tables are out there. If you're going to have art outside as part of a creative placemaking effort, the art needs to be made with the community and needs to connect with them in real, hands-on ways. Doing art about the community or doing art in front of the community doesn't help undo the social damage we've caused in the way we've designed and created divided, auto-centric cities; in the ways we've segregated people; in the ways we've ignored the importance of equity, opportunity, and social connectivity. Like Rick Lowe figured out as he started working on Project Row Houses[4] in Houston or as Frances Whitehead[5] considered as she moved to 'post normal' art in Chicago, why make art about the challenges of our times when we can make art that at least begins to take on things directly? Why not make art that actively intervenes?

MARSH: Placemaking, really, is about re-evaluating the world and how it is changing. Everything we create is ephemeral. It's an illusion to believe that a project or place people use is ever going to be finished or used in the way you plan. With urban design, public art, architecture, and landscaping, people who use these will show you if your intended idea for a space is right or not. America is a relatively young country. When you look at buildings and public places here, many have only been in existence a hundred years; in rare cases, a couple hundred years. Then you compare that to other parts of the world where you have buildings and communities in existence for several centuries. There's a lot to learn in how places are used over time. And places constantly need to be re-evaluated to ensure they're meeting the desires and needs of the changing culture of the people living there or visiting.

WALKER: Yes. Placemaking is a response to this re-evaluation. It's an action step addressing changes. Placemaking works best, at least right now, as place repair. So what the thing people are doing, at least most of the time, with placemaking today is trying to undo the problems created by suburbanization and breakdowns of social networks. We have re-evaluated those choices and believe we need to do something about the damage. We're taking a tactical approach to place or city repair – also the excellent name of a group doing this work for decades in Portland, Oregon.[6] With the way sprawl spread out cities, especially in America, places don't have the population density to give the right level of life and activity to places that once were inherently vibrant and sticky – that were places where people stuck around and spent time together.

In the past, many cities and neighborhoods and blocks – especially with a high density of residents and streets that remained safe and comfortable for people – didn't really need public programming or placemaking. When it came to street life or pedestrian activity, this all just worked – even while other problems, especially on the social level, still existed. And many places around the world still work will without placemaking interventions.

Earlier in this trip, we saw a big difference when comparing the market area in Pittsburgh – called Strip District[7] – and the Kensington Market and Chinatown market area in Toronto. The Strip District is a really wonderful place, but it goes really quiet in the evening and at night because all the people go other places, many back to the suburbs. In Toronto, the density of people living and working in Kensington Market make it so much more alive twenty-four hours a day (Figure 8.1).

MARSH: So the purpose of placemaking, many times, is to attract more people, via unique experiences, to either visit or move people back into these areas that previously did have that higher density of people. Those places like Kensington Market don't even have to think about programming. It's already that way. And it's not always like these denser cities have great architecture or design. They just have a lot of people. The people take over and you see a lot of graffiti and people doing what they want to do outdoors. A place with high density of people may not require placemaking for the purpose of attracting people to public places. But active placemaking programs, like socially engaged art projects, can be used to connect people, connect strangers. This is really important given the current social climate where we are dealing issues of equity and political polarization. Artists and placemakers creating social opportunities for people of different backgrounds, with different beliefs, can help support empathy and understanding. This can be beneficial to creating a happier, more harmonious city. People start seeing each other as individuals and not as somebody from a different tribe when they play chess or ping pong with a stranger who may think differently. Placemaking is providing that opportunity.

WALKER: Absolutely. The advantages of connecting people with each other should be a primary goal of this work. And I think the work should also have long-term goals focused on reducing the role of the automobile in our lives. People should be able to enjoy public places very near where they live. And we should live and shop and work and find entertainment and recreation in close proximity as much as possible. A life with less driving and more walking and public transportation is better for physical and mental health, keeps businesses like corner shops and pubs going, and helps with empathy. When you are out on the sidewalk or in a packed train, you have to deal with other humans in real space and time and you get out of your bubble. This is more important than ever, especially here in America.

The Strip District in Pittsburgh could use a lot more people, especially at night. It seems like a great place to live and residential development appears to be coming there. Placemaking and public programming could help make that place even more appealing and push this forward. When you see the Strip District, you see a place that's perfect for people. So the problem isn't the place, it's the overall design of the city that prioritized getting in and out of the city, not

What do we believe?

Places are made when they feel like home.

When everybody is welcome and comfortable.

Artists can and should support life in neighborhoods through culture, creativity, and community.

Artists are good at home and place making.

That's why we do the work we do.

What is our work?

We support a better quality of life for people by utilizing the tools of art and culture to encourage us all to be more...

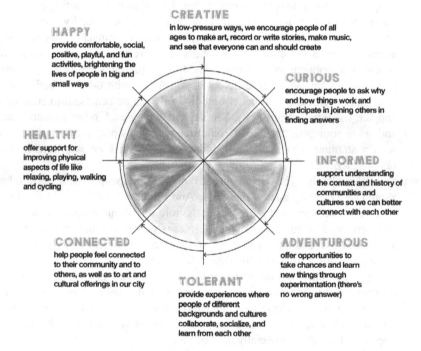

CREATIVE
in low-pressure ways, we encourage people of all ages to make art, record or write stories, make music, and see that everyone can and should create

HAPPY
provide comfortable, social, positive, playful, and fun activities, brightening the lives of people in big and small ways

CURIOUS
encourage people to ask why and how things work and participate in joining others in finding answers

HEALTHY
offer support for improving physical aspects of life like relaxing, playing, walking and cycling

INFORMED
support understanding the context and history of communities and cultures so we can better connect with each other

CONNECTED
help people feel connected to their community and to others, as well as to art and cultural offerings in our city

ADVENTUROUS
offer opportunities to take chances and learn new things through experimentation (there's no wrong answer)

TOLERANT
provide experiences where people of different backgrounds and cultures collaborate, socialize, and learn from each other

Figure 8.1 What is our work: Big Car Collaborative takes a holistic approach to its work, seeing placemaking as a strategy for reaching important, broader goals with communities.

staying in it. Like most larger American cities, Pittsburgh is disrupted in major ways by downtown freeways. That's hard to fix. And public programming can't do much about that situation. But the way downtown and public areas are programmed can change. And placemaking can be part of that.

When we were in Buffalo, you could see that they have been working on their downtown – especially their riverfront area – in a way to primarily make it appealing to people who live in the suburbs and drive in for big, self-contained festivals and events. They've dedicated lots of space to parking and they've made it easy to drive in and drive out. These kinds of ticketed festival grounds are more self-contained, inwardly focused islands where everybody spends their money inside the fence. And, because of this setup, the events don't support small businesses nearby. So you'll see very little life in a downtown bar three blocks away while people stand in line to buy a beer at the festival grounds. It doesn't help the businesses and other places around it. Everybody buys their beers and food in the festival. And it doesn't have any sort of economic development support for those depressed blocks right there by it, where there's a dead shopping mall and other vacant spaces. There were hardly any people down at a really great bar at the Lafayette Hotel – just blocks from the festival full of thousands of people. That's the problem. And we have the same issue in Indianapolis. You park in the parking garage, go to the White River State Park,[8] go to a concert. Then you never go to anything else. You don't really touch the experience of being in the city or what it is about. It wouldn't matter where the festival was located, the festival gives a person no experience of the city core, it just happens to be in the city. You might as well have it in an amphitheater in the middle of a cornfield. It's the same experience.

MARSH: A lot of the people working for the city, creating programs for the downtown core, live in the suburbs. I think that has something to do with it. What do you think can be done by cities to keep people around the core?

WALKER: Cities should be designing their downtowns and planning their cities for residents to live right there. And public programming and placemaking work should be designed to include the people who live there – the homeless, condo owners, service industry and office workers, CEOs, everybody. There are already communities in these downtowns. Focusing on public programming like festivals and concerts that often exclude people because of cost ignores a good portion of these community members. You often see them there outside the fences. Instead of putting all of your eggs in that blowout event basket that many can't afford, why not offer lower-cost, human-scale public programming more often? We've seen this work really well in a lot of places from Bryant Park[9] in New York to Campus Martius[10] in Detroit to Monument Circle in Indianapolis.

Of course, Indianapolis has messed up a lot with that over the years as well. We've built our city and prioritized our streets and roads to serve people who live in the suburbs. And we've done the same thing, far too often, with events designed to attract suburbanites. If you want people living in the core of the city, you have to design the downtown and your public programming for people who live there, who would live there, and not for people who live somewhere else. If you keep cities and places designed in a way that makes it too easy to live somewhere else, people will keep on living there.

But if you design public places in outwardly oriented ways, they'll help support what's around them. Look at Chicago's Millennium Park.[11] It has attractions, a parking garage people from the suburbs drive into, big festivals. But it's also designed for people to hang out in there for free, experience art, and socialize. And it doesn't gobble up all the business from around it. So people can go to Millennium Park but they also walk right into the city. It's about integrating the design of the public place into the city. It's not separate, not fenced off. Bryant Park in New York is like that too. They made it part of the city. You might buy something – like a drink at the sidewalk café – at one of those places. But you're going to move on to the next thing. That's where placemaking can help. It encourages people to go on to the next thing, to wander and to have spontaneous experiences they wouldn't have if they were trapped in a programmed festival.

MARSH: Like when we were doing *Spark*[12] at Monument Circle in the center of Downtown Indianapolis. The owner of the jewelry store and owners of some of the other businesses said when big festivals pack thousands of people onto the Circle, they actually get less business than when nothing is happening down there. No one goes into their businesses because they are only there for the big, loud festival parked in front of their business. But they said that *Spark* – which created opportunities for people outside and experience quieter, human-scale activities – encouraged people to stay around and actually support the businesses. People were moving from place to place and experiencing the city. They didn't just do the one big thing, blow it all out, then drive back home to the suburbs again. That's how cities and how neighborhoods work. Cities weren't supposed to be a place where you went to just one thing then turned around and drove twenty miles back home. You'd go from the corner market to get fruit, then you'd go to the other corner and get meat, and then you'd go down the way and get some bread and cheese. Then maybe stop in the bookstore. And you'd run into people along the way and have conversations with strangers and maybe make new friends. Now, shopping happens much of the time in one, big place. And life is all split up and the in-between experience is isolated inside cars. If the supermarket has everything you need, then you don't go into the different shops. And you lose out on a lot of opportunities for human interaction, for meeting your neighbor, meeting the different shop owners.

Some of this still happens when you see your neighbors at the Walmart in a small town. But it's less likely that you'll have spontaneous interactions at these big box stores. Also the problem is how you get to Walmart and get home, is in your car. You don't see anyone on the street. If you're in Toronto and in Kensington Market and getting three or four things you need you're going to run into more people and talk to them.

WALKER: In Indianapolis, when we go down Virginia Avenue by Bluebeard[13] and Calvin Fletcher[14] coffeehouse, it's like Mayberry, from *The Andy Griffith Show* on TV back in the '60s. You walk down the street and see friends and catch up. It takes time but it is important. That's a lot more like an authentic experience. Just like we saw at La Prima[15] – this Italian espresso shop next to an Italian bakery in Pittsburgh. There are plastic chairs out front under a large awning shared by the two shops. Customers, some speaking Italian, just show up there and meet up and form friendships. It's kind of like how I used to play pickup basketball. We'd show up individually and do this group sports thing. There, it's a group cup of coffee. We heard them telling everybody that they'd be having a celebration the next morning for somebody moving away. They were acquaintances hanging out at these plastic tables and chairs they pull together, having coffee and pastries together and talking and celebrating. Sometimes the customers are also sitting there playing cards. La Prima is a made place. It doesn't need city repair, it doesn't need anybody's interventions because it is still working. It has great design: a big, metal awning to protect you from the elements and movable furniture where you can sit and watch people go by in the Strip District for as long as you'd like. It's not designed, like many fast-food restaurants, to encourage you to move on. La Prima offers excellent food and drink right there by the sidewalk. And the customers have created their own avenues for social interaction.

People don't get that kind of experience at Walmart in big cities. They miss out on these routines, on seeing people on the sidewalk and the street. We have designed our cities for efficiency but that took away a lot of the social spontaneous stuff that made life really good. And, as we've seen, sprawl is very inefficient in reality. It's turned out that people spend all the time sitting in traffic and waiting in lines that they once had for conversations with neighbors. Working on repairing these bad choices is where placemaking and an art practice to encourage social interaction can have a positive effect. Maybe we can be part of helping fix problems that were created by capitalism and greed and always thinking things need to be changed and improved when really it was just destroying things that are already good. Changing things for the worse, for the illusion of progress. These changes – like the incessant and unnecessary road building that continues today – are all linked back to somebody making a lot of money. But these are short-term profits that contribute to long-term problems.

MARSH: What do you think people want? Do you think it's different from city to city, is it unique from city to city or is there a formula?

WALKER: People want to be around other people. They want to have opportunities to spend time together or to be comfortable, feel safe. A lot of those are basic things. And many public places don't allow for that kind of stuff. Some of this is because people in charge of public places don't want you to get too comfortable or you might move in. We're seeing that in the park in our neighborhood as homeless folks have set up camp. But when you make benches purposefully uncomfortable so drunk or homeless people won't sleep on them, then you are also making them uncomfortable for the rest of us. And it tells us all the same message that was meant for the homeless: 'We don't really want you here, so move along.' (Figure 8.2)

MARSH: In America, we've built a society where people stay inside much of the time – often in front of screens. Add driving into the mix and we are very physically isolated. I don't think that's what people really want. We like to move around and do things. But going to a public place and connecting with others is now such an ordeal because of the way things are separated. It goes back to what we were talking about with Buffalo. People shouldn't have to buy a ticket, drive fifteen miles downtown, park, and then go to the festival grounds to find an opportunity to connect with others in a public place. This kind of experience needs to be something that is an easy, natural part of life. That's where designing something to add a little pause in a public place is important – like we saw with a Southwest Airlines and Project for Public Spaces[16] Heart of the Community project outside the public library in down-town Buffalo.[17] The idea of small interventions in public places, in the case of the library turning a modernist plaza into a gathering place, is good and will make a difference for all kinds of people. We've seen this same thing with our work with Indianapolis City Market[18] – people are there for a variety of reasons because it is by the City County Building and the courts. So we've added features that are comfortable and respectful to everyone. Maybe you are the mayor or maybe you are there for a court hearing.

WALKER: As we look at this, I see the small things as important and as things we can do now. But, in the meantime, we can't just wait for the big things to happen. Policies need to be influenced to encourage changes in how cities and regions work. We know this is failing. But we still haven't been able to make sure that we stop further suburbanizing and lowering the density of our cities. Without density returning to urban neighborhoods in cities like Indianapolis, we'll continue to struggle to enjoy the kind of thriving public places and neighborhood commercial and cultural corridors we see in very dense cities like Toronto.

How much do you see placemaking and socially engaged art making a difference?

MARSH: It varies from city to city. Placemaking has always been around. And now it has a label. This is because isolation and tribalism and nationalism have become such major issues. What's happened was the world became so small via the internet but also became so isolated also due to the internet and technology. It's complicated because, with the internet, our species has created a tool for creating world-scale tribes. In the past, we created groups based on geography. The internet has changed geography. Humans are social. We seek connection. Placemaking creates opportunities for social interactions with

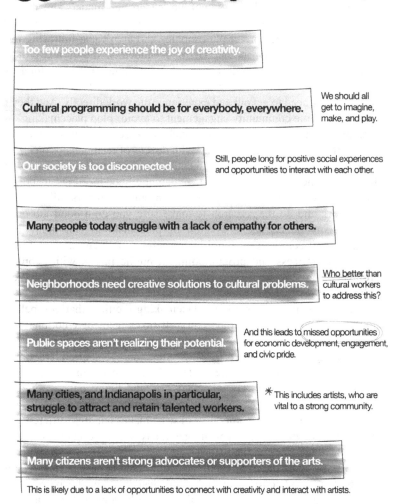

Why is this work so important?

Too few people experience the joy of creativity.

Cultural programming should be for everybody, everywhere.

We should all get to imagine, make, and play.

Our society is too disconnected.

Still, people long for positive social experiences and opportunities to interact with each other.

Many people today struggle with a lack of empathy for others.

Neighborhoods need creative solutions to cultural problems.

Who better than cultural workers to address this?

Public spaces aren't realizing their potential.

And this leads to missed opportunities for economic development, engagement, and civic pride.

Many cities, and Indianapolis in particular, struggle to attract and retain talented workers.

※ This includes artists, who are vital to a strong community.

Many citizens aren't strong advocates or supporters of the arts.

This is likely due to a lack of opportunities to connect with creativity and interact with artists.

Figure 8.2 Why is this work so important: It's all about the 'why' or the motives behind placemaking in the view of Big Car Collaborative's team. Putting it in words helps clarify and guide the work.

people different than you and helps you understand people who are different from you. It provides that opportunity for unplanned interaction. With your social media account, you create like-minded groups. We need that. But, to improve our future and our lives, we need interactions with people much different than us. Placemaking should be funded more and more by cities and by foundations. We are already seeing this in cities with smart leadership. Indianapolis is a place where some of our leaders and funders already get it.

However, funders should be making sure artists and organizations pitching placemaking have the correct philosophies and aren't just following money. We've seen it in Indianapolis. All of a sudden, a lot of non-profits are placemaking. Some are doing a fine job of it, others not. The foundations shouldn't rely on grant reports or videos. They need to visit their funded projects. They need to get out and experience the work directly. And they need to emphasize true community engagement to avoid 'plop placemaking'.

People don't always realize how much work goes into this work. We are really in the people business, dealing with dozens of relationships on all levels. This means going to neighborhood meetings and being an active part of these. Meeting with neighbors who don't go to neighborhood meetings, meeting with businesses owners and teaming up with them and neighbors alike to lay the groundwork and do the research to make sure you do a good job on these projects. Anyone can do a version of placemaking. But doing it in a thoughtful, effective way requires research, practice, and staffing the engagement process and programming in public places with people who care. These have to be people who aren't just doing this for a paycheck. We're fortunate to have found many artists, designers, and planners who share this passion with us. Good placemaking requires that you believe in what you're doing and its importance. It takes a lot of emotional energy to be a bridge for people, to connect people and create spaces and programming where comfort, play, fun, and social interaction can happen – all at no cost for the visitors. You have to smile at people and say hello, ask how they are, who they are, and figure out the situation.

You can put some games out or set up your art studio on a sidewalk and call it placemaking. But a key component of what makes our style different is your journalism background, Jim. You are genuinely interested in the people in the places where we work, in their stories.

WALKER: Yes, the work has to be about the people and not the placemaker. For the most part in good journalism and good design alike, the person who is making the story or the space doesn't need to be apparent in what's going on. When they are invisible, that means they are doing things right. It's like a referee for a basketball game. You know they did a good job if nobody is paying any attention to them. If the game starts to be about the ref, something went wrong. In placemaking and socially engaged art alike, the designers of the space, the project, and the programming should be responding to what people want. And they should be doing what they can to make that happen – usually in behind-the-scenes, invisible ways. With this work, I still think about what I want as a basic test. Anything we offer has to be something I'd want to do or a place where I'd want to hang out myself – and I always spend hours and hours of time in these spaces, doing the activities with people and making adjustments and improvements. But these projects are about the people who participate in them, not about me or us.

MARSH: The really difficult, delicate work is with people – facilitating interactions between people. It helps to have a multi-pronged approach. Public placemaking is more on the front lines connecting people, removing as many barriers as possible to get people to connect and permission to be creative. Because of social structures our society has created, people don't feel like they should make things, either because they didn't go to art school or because they can't make a profit at it. But there's a benefit to people to create and interact with each other through art.

Often people happen upon our public placemaking work, then want more opportunities. That's why we have our home base, *Tube Factory*,[19] operate equally as a community center and contemporary art museum. That's why we chose to give 2,000 square feet to a place where neighbors can have meetings and gather and 2,000 square feet for the gallery where we commission shows curated on the themes of mythology, community, and memory. When selecting artists and exhibits, I look for ways to bring different audiences together. With the artist Carlos Rolón,[20] we presented *50 GRAND* (2017),[21] a standard 2D art exhibit and an installation. He also created a boxing robe that members of a local boxing group could wear. And we brought in a full-size boxing ring and hosted free live boxing matches in partnership with a boxing gym that serves at-risk youth. Rolón is a well-known artist who would also attract an arts audience. So we were able to bring together two different audiences with that show that would not have many options to interact typically.

That's why I selected those themes. People of different backgrounds have things in common, certain childhood foods, smells, tastes, décor from grandparents' and parents' homes. I learned about theme-based curation initially at Tasmania's Museum of Old and New (MONA).[22] They curate based upon the themes of sex and death. Chicago artist Theaster Gates[23] and MONA are two of my biggest influences.

WALKER: I also really enjoyed another combination of audiences that happened with our *Hairy Man* exhibit you organized with Jeremy Efroymson[24] and Christopher Murphy. The public programming included talks by experts on Bigfoot. The Hairy Man[25] is the Native American folklore version of Bigfoot and the show explored this and the obsession with this idea. So here we were, just a few months after Donald Trump was elected, hosting rural Bigfoot hunters from across the Midwest and artists from the city together in this space, sharing paranormal theories. The highlight came when the guy who hosts the Ohio Bigfoot Conference told the audience they should support art spaces like ours and artists in their own communities whenever they visit a new place. This kind of moment really felt like it was bringing people together.

The biggest thing I want to do with placemaking, socially engaged art, and tactical urbanism is create these kinds of small opportunities and interventions that are part of bigger, long-term societal and policy change. We have to do something right now to bring people together in our world. The goal needs to be an emphasis on how this work can contribute to empathy, equity, better health, safer communities, places where people are physically and socially connected with each other and the rest of their city. And we need to keep in mind the potential negatives of gentrification, especially displacement of people from the best parts of cities – neighborhoods close to jobs and parks and art and good food and all the things that make places attractive. The work of creative placemaking needs to be for everyone. And we shouldn't be making places better just for developers to make more money. So it's important that people don't feel satisfied with placemaking as window dressing or as a tool for the kind of economic development that benefits only a few. We need to be sure we are positively influencing the ways cities are changing for the benefit of many.

MARSH: Do you think there's incentive for people to be less passive about these challenges?

WALKER: We need to lean on the influencers – the developers, the funders, the city leaders. On a small scale, we can do things that help people become more active within their own communities. But, on a larger scale, we need to continue to get city leaders and elected officials involved in these ideas – in supporting a better-connected society and safe, comfortable, equitable, artful, and welcoming neighborhoods and public places. There are two things we have to do: One is take on these smaller interventions, team up with neighborhoods, with all kinds of people, and just make things happen out there. Then we also have to do bigger things that influence politicians and funders to get behind this on a policy level by helping them see, experience, and believe in the power of this work.

Notes

1   Big Car Collaborative, www.bigcar.org
2   Fountain Square, www.discoverfountainsquare.com
3   National Endowment for the Arts, www.arts.gov
4   Project Row Houses, www.projectrowhouses.org
5   Frances Whitehead, www.franceswhitehead.com
6   City Repair, www.cityrepair.org
7   Strip District, www.stripdistrictneighbors.com
8   White River State Park, www.whiteriverstatepark.org
9   Bryant Park, www.bryantpark.org
10 Campus Martius, www.downtowndetroitparks.com/parks/Campus-Martius
11 Millennium Park, Chicago, https://www.cityofchicago.org/city/en/depts/dca/supp_info/millennium_park.html
12 *Spark* Monument Circle, www.bigcar.org/project/spark
13 Bluebeard, www.bluebeardindy.com
14 Calvin Fletcher Coffee, www.cfcoffeecompany.com

15 La Prima, www.laprima.com
16 Project for Public Spaces, www.pps.org
17 Heart of the Community, www.pps.org/blog/2017-heart-of-the-community-placemak
   ing-grants/
18 *Spark* City Market, www.bigcar.org/project/citymarket
19 *Tube Factory*, www.tubefactory.org
20 Carlos Rolón, www.carlosrolondzine.com
21 50 GRAND, www.bigcar.org/project/50grand
22 Museum of Old and New, www.mona.net.au
23 Theaster Gates, www.theastergates.com
24 Jeremy Efroymson, www.jeremyefroymson.net
25 The Hairy Man, www.bigcar.org/project/hairyman

Reference

Lydon, M., Garcia, A. (2015). *Tactical Urbanism*. Washington: Island Press.

Section 4

Challenging Ecologies

9 Temporary spatial object/architecture as a typology for placemaking

Torange Khonsari

Abstract

This chapter will introduce the role temporary spatial objects/architectures can play in claiming space in parts of neighbourhoods for both civic use and social empowerment harnessed towards placemaking. Research within creative place-making that considers contributions to citizen-led creative placemaking is emerging and this chapter argues for a tactical approach to creative placemaking. Temporary spatial objects/architectures are presented as a tactic to encourage citizen-led approaches to resist privatised enclosures in urban areas. This chapter will use two historical contexts of temporary spatial objects/architectures embedded in fields of architecture, site specific performance and art; the Soviet Agitational Propaganda Vehicles (Agitprop trains) of the 1920s and the Fun Palace by Joan Littlewood and Cedric Price of 1960s. Two selected London case studies of temporary spatial objects/architectures by public works [sic] collective will offer the contemporary position of such structures within the neoliberal neighbourhood of London. This chapter argues for temporary spatial objects/ architectures to act as what Flood and Grindon (2017) call 'disobedient objects' in placemaking projects that battle against waves of city-development, supported by capital interests moving at a high speed. To this end this chapter explores how temporary object/architecture in its disobedience creates agency within a locality encouraging chance encounters and organic formation of communities.

Introduction

This chapter will primarily draw from two case study projects by public works collective:[1] firstly, a mobile structure, *DIY Regeneration* (2009); and secondly, a situated temporary project, *The Bow Common* and *Common Room* (2014–ongoing) as possible objects of disobedience. The case studies' context is London, an ever-growing city with increasing gentrification, where the high value of land prohibits placemaking tactics described in this chapter. It will draw comparisons with two historical case studies, which share commonalities and differences. For mobility, it will look to use the Bolsheviks Agitprop trains of 1920s, in Russia, and for temporary architecture it will draw on the adaptable *Fun Palace* of theatre director

Joan Littlewood and architect Cedric Price of the 1960s in London. Both the historical and the contemporary case studies share the commonality of being situated in a place to affect a social impact and are offered as current examples on the historical trajectories of the past. The contemporary structures are contingent within a complex and unpredictable city where residents have moved away and new ones have moved in: 'Complete participation would suggest a closed domain of knowledge or collective practice for which there might be measurable degree of "aquisition" by newcomers' (Lave and Wenger, n.d., p.36). Every move brings with it new cultures, desires and values which the ephemeral nature of these structures can accommodate. The term 'spatial objects' refers to mobile structures created by artists, architects and designers, whereas 'temporary architectures' are structures created by the same group of practitioners but are fixed to a site and are considered temporary by planning authorities.

public works collective was set up in 2004 as an interdisciplinary platform between artists and architects situating themselves in neighbourhoods using temporary spatial objects/architectures as a method for social engagement. The methodology used in this research is situated practice in both case studies, being in residence as a spatial practitioner through the collective, public works or as an academic through London Metropolitan University. The act of being situated in a place where relationships are made, collective tactics agreed on, and being involved in events, allowed the researcher to become part of the community of learners attached to a locality and place (Lave and Wenger, n.d.). Here local knowledge was produced, disseminated and acted upon. Most of the findings in this chapter are based on experienced knowledge within the communities of learners. These findings were then analysed against desktop research of the two historical projects, which have commonalities and conflicts with the contemporary case studies.

This chapter will not elaborate on specific tactical methods, but positions the temporary spatial objects/architectures as a domain for tactical knowledge to be shared and tactical moves agreed upon. It is from within these structures that tactical moves are devised and acted on. Temporary spatial objects/architectures, although highly programmed with events and activities, are not about being purely functional. They are concerned with the nurturing, gathering and dissemination of knowledge, collective decision-making and the offering of platforms for individual desires, for both individual and collective voices to be heard. By their placement on to sites in the city they have the potential to become agonistic in nature (Mouffe et al., 2001), a space where conflicts are negotiated and agreements are made. They require non-corporate public spaces, where pluralism (Mouffe, 1992) can be hosted within the theatre of everyday life away from corporate control. When temporary spatial objects/architectures become the arena where civil society claims its right over the use and possibly the ownership of places, they become what Flood and Grindon (2017) call 'disobedient objects'. The more these structures can create their own financial independence, the more they can respond to its role as places where democracy is practised. This chapter takes its understanding of democracy from Mouffe's conceptualisation, where democracy becomes negotiations among interests, rather than those scholars who think in lines of consensus and pure 'communitarian' spirit (Mouffe, 1992, p.29). In the current social context in the UK, where the social value system is so wide and varied, we need places that can embody and understand such variety. It is important not to simplify and order it with imposed values, which in turn requires what Chomsky and Herman (1988) term 'manufactured consensus'.

Situating spatial objects/architectures in place

Based on knowledge gained through being situated both in Bow and Kings Cross in London, temporary spatial objects/architectures aiming to make places are programmed with themes derived from local involvement of citizens. The structures are offered as open platforms for local initiatives generated at the grassroots and arenas where tactical or specialised local knowledge is exchanged, aiming to engender agency in those involved. It is these spatial objects/architecture's embedded nature away from financial systems of consumption that aligns this typology with 'situated' environments for learning for communities of practice (Lave and Wenger, n.d.). These spaces are not about being commercially productive but are set up to nurture collective action. They are places where knowledge for one's right over the city can be learnt and exercised as citizens; here, citizenship is practised. As Mouffe et al. tell us:

> Arendt, when she speaks of the right to have rights, insists that the right to have rights is citizenship. If you are not a citizen, you don't have the right to have rights. Thus, in fact, she insists very much on the importance of being a citizen.
>
> Mouffe et al. (2001, p.105)

If we see citizenship based on pluralism with an understanding that at a higher level we all belong 'to a political community whose rules we have to accept' (ibid., p.30), it is pertinent to question the structure of the localised political community. Who makes up the political community, the rules and its governance? Temporary spatial objects/architectures 'situated' in a place have the possibility to test and question relations of power, be it top-down or at grassroots. The testing can be through models of platforms offered for discussion, thematic of workshops and events and involvement of expertise offered by local residents. This can influence localised governance of neighbourhoods and their resources, decision-making on ways in which sites get developed, methods in which commonalities and differences are celebrated and negotiated and visual representation of values displayed in festivals and public events.

Temporary spatial objects/architectures in their disobedient form are structures whose aesthetic, flexibility and affordability enables their multiple use and misuse, and through this process allows them to remain contingent. The motive force and the affordable way in which the disobedient object/architecture can establish itself gives it power – the power of being possible, present, constant and 'extending material spaces into spaces for action' (Arlt et al., 2006, p.14).

Haydn discussed material as being generated through human relations and programmes in flux, away from current obsession with efficiency and order, prohibiting chance encounters and organic formation of communities. These spatial objects/architectures growing out of interdisciplinary praxis often have diverse materiality ranging from:

> as found, originating from the French object trouvé to being products of manufacturing (architects tend to lean towards this due to disciplinary education). Some structures will be made to enable ease of collective construction demonstrating social solidarity, whilst having the ability to change and shift with time. Disobedient structures can exert 'counter power'.
>
> (ibid., p.9)

Two key constitutive components of such structures are firstly, delivery of social agency, and secondly, re-imagined models of public spaces towards active citizenship. To implement these constitutive components, projects need to be self-initiated with new models of financing which are dependent on countries' economic cultures, values and policies. These can range from grants to crowd-funding, as well as becoming income-generating. Mobility presents a different relationship to being 'situated': it is situated in shorter periods and its role has less impact than temporary architecture in creating long-term agency. However, it contributes in producing new models of public space in places it visits, as described in the next section.

Temporary spatial objects

Historical context 1: Agitational Propaganda trains (Agitprop)

In order to frame the discussion on spatial objects, this chapter introduces the historical context of Agitational Propaganda trains, referred to as Agitprop. Here I am not claiming this is the primary historical context to subsequent development in the art, architecture and design of mobile structures; rather, I am exemplifying it as a precedent, which had political intent and whose intensions were fulfilled using mobility of a found object (a train), married with contemporary constructivist art of the time.

The Department of Agitation and Propaganda was founded in 1920 by the Soviet Communist party. Russia's population was 70 per cent illiterate, thus visual language of art and theatre was the most effective means of spreading revolutionary ideas (Brown, 2013, p.5). The Agitprop trains were not passenger trains but carriages utilised on the national railway tracks to reach as much of the country as possible, becoming a messenger of utopian dreams, where constructivist film makers such as Alexandr Medvedkin, and artists such as Malevich and Lissitzky, produced artistic visual content, creating the theatrical spectacle to lure audiences. Imagine the scene as trains approach the platform, hordes of farmers with hoes, shovels and sickles run towards the station. The train throws poems, manuscripts and flyers out of the windows. Among the revolutionary flyers and the intense red posters lie helpful tips about farming, repairing your house and bicycles. The train stops, doors open and a film begins, the library opens and a member of the Blue Blouse, an influential Agitprop theatre collective in the early Soviet Union in 1920s, stands on the roof performing revolutionary messages. Utilising Agitprop, art, theatre and poetry was being developed hand in hand with top-down political ideology. One can question if this set the historical precedence for what UK Prime minister Tony Blair in 1990s advocated in policies towards public art in regeneration, which subsequently enabled art practices to develop in the public realm funded by the state, questioning, as above, self-initiation and the role of multiple funding sources to achieve artistic freedom within the public realm.

The instrumentalisation of art of Agitprop with top-down political ideology becomes a cautionary tale. One can take the view that the emerging Russian government invested in new forms of theatre, methods of art production and aesthetic, promoting a synthesis between many visual and political re-enactments into the realm of art and culture, which was important for artistic development of the time. Agitprop in bed with art and culture of the time in Bolshevik Russia is both fascinating and disturbing. Brown quotes Pipes on Agitprop, describing it as a rise in 'cultural bureaucracy for whom culture was only a form of propaganda, and propaganda the highest form of culture' (Brown, 2013, p.6). Working classes had no time or money to venture to cities to attend theatres so the trains went to them. There was a lot of improvisation, varied engagement with audiences and active participation to really engage people, to keep them interested and keen to attend other performances. The creativity in social engagement, championed in the 21st century through abolishing the audience/performer/artist dialectic, and the notion of useful art, were all emerging. In this sense constructivist art around the Agitprop phenomenon becomes an important reference point in current socially engaged art practice, participatory theatre and art in the public realm, be it mobile or static. The cautionary tale of this type of instrumentalisation of art by political power is the implementation of didactic propaganda messages it spread using spectacle and simplified, easily accessible content. For the artists the ideology of empowerment of the working classes against oligarchy was more important than power relations and manipulation being set up by communism and their political strategies. The constructivists and their Agitprop projects set the scene for interdisciplinary endeavours post the onset of industrial revolution.

Case study 1: DIY Regeneration

In 2009, Camden Art Centre commissioned public works to do a project in the Kings Cross redevelopment area, one of the biggest regeneration schemes in Europe. The project, *DIY Regeneration* (2009),[2] used the *Folk Float* (initially commissioned by Grizedale Art centre),[3] a customised milk float, as a mobile workshop space, billboard, archive and an on-site office. The milk float and public works roamed the area looking for activist residents, workers and commuters engaged in making a difference to the area. *DIY Regeneration* was interested in small-scale, self-driven initiatives, motivated by local actors as well as local needs and desires. Over the three months, public works talked to individuals who had a direct, active involvement with their community and on a daily basis contributed to and changed the area in which they lived. Each of these chance encounters at the *Folk Float* which resulted in handmade posters captured the tips and advice offered for others to get involved and start taking ownership of public spaces in one's neighbourhood.

The content was not produced from the top as a means of dictating values, like the Agitprop, but was derived from knowledge of everyday life in a place. The posters, which documented the tips, advice and slogans collected on site, were displayed on the roaming *Folk Float* as well as on an online specially designed digital platform. A final selection of posters was printed and fly-posted across the Kings Cross area, spreading the advice back into the community which produced it. Returning to the cautionary tale of the Agitprop, the spatial object developed by public works was not a means to make government consultation events or a method to make government agendas come to life. For this their value came from the local neighbourhood and not local government or the developers. The structures were meant to reveal hidden informal sociabilities to allow collective interests to reveal themselves. Here is where the unexpected is revealed, and knowledge required for civic action disseminated towards the heterogeneous city: 'I'm trying to think of a model of the public sphere which will not be one where people leave aside all their differences in order to try to reach a consensus, but precisely a sphere where an agonistic confrontation takes place' (Mouffe et al., 2001, p.123).

In *DIY Regeneration*, residents' frustrations, views and disempowerment were revealed, shared and discussed, yet the ephemerality of the structure in a place did not enable lasting relations that led to trust and potential agency. Mobility is both a vice and a virtue, it is very short-term presence in a place does not achieve commitment for setting up a true agonistic platform, from which to empower, mobilise civic action and support self-governance. It is more a tool for spreading words/knowledge and campaigning like the Agitprop. Mobile structures are effective for reaching otherwise difficult communities with low participation. Used to benefit civil society, their agency lies in bringing knowledge and a different value system to a place, and these differing values have the potential to create agitation and what Mouffe calls agonistic (Mouffe et al., 2001) places of difference. With Agitprop the agitation was part of a political strategy to change public opinion; this is different to *DIY Regeneration*, which offered knowledge on a local scale for citizens to use tactically, sometimes in opposition to strategies imposed. In this sense they may become disobedient, with the potential to counter power exerted by capital and legislation. If the practitioner entering areas anticipated for development starts with deploying mobile structures, both their strategic and tactical intent needs to be thought through and collectively discussed and agreed on by local residents. In this context the mobile structure becomes a space for productive social engagement and where knowledge is transferred.

Temporary architectures

Historical context 2: **The Fun Palace**

Less than half a century after the Agitprop vehicles of the Soviets and half a century before *DIY Regeneration*, Joan Littlewood engaged in revolutionary Agitprop street performances in the UK. Littlewood's work was to occupy the spaces of the city, the streets and the neighbourhoods, her passion shared by architect Cedric Price who saw architecture as in constant flux, and temporary. Price and Littlewood's unrealised project, *The Fun Palace*, was reflective of its political context: *The Fun Palace* was to be an educational and recreational centre in which users would 'actively participate, instead of passively receiving entertainment' (Shubert, 2005, p.5). It was to have a multitude of activities, from music, performance and theatre to hangouts. Littlewood and Price saw the spaces as varied in size, lighting and shape. Price saw the space as something that never reached completion and was to host informality: 'The programmatic fluidity and formal indeterminacy of *The Fun Palace* might be thought of as an architectural analogue to the transformations experienced throughout post-war British society' (Mathews, 2005, p.91).

The political context within which Littlewood and Price were operating was filled with social ambition. The Labour leader at the time, Aneurin Bevan, had initiated welfare state reforms around leisure and making higher education available to all classes. With a decline in industry, the need for unskilled labour decreased and the working class had more time to call their own. Leisure was a key political and economic agenda, which the government was promoting with the perception of moving the working classes away from drinking and crime and revolutionary ideas towards more consumerist leisure (Mathews, 2005). All classes were admitted to university to learn more intellectually challenging subjects, many in the leisure industry and tourism. Price believed people should be free to find their potential, interest and skills in an environment such as *The Fun Palace*. Mathews (2005, p.79) tells us that 'Price thought of *The Fun Palace* in terms of process, as events in time rather than objects in space, and embraced indeterminacy as a core design principle.' The strength of *The Fun Palace* for temporary architecture as agents for change was emerging in Price's ideas with improvised architecture and Littlewood with undetermined and interdisciplinary programming. Their collaboration with Gordon Pask and his interest in games theory within *The Fun Palace* leaned towards negative social engineering of the Agitprop. Cybernetics, interactivity and social control took the place of Bretchian theatre and Price and Littlewood, in the hope that human control could be diminished, allowed this to happen (Mathews, 2005). The complex and rich ideas in *The Fun Palace* has enabled its lasting influence, which may not have been as powerful if it had ever been built and its social engineering elements realised.

The Fun Palace's idealisation of human behaviour is avoided by the situated condition of this type of temporary object/architecture. Being situated enables understanding of human behaviour to be concrete rather than imagined. *The Fun Palace* happened at a time where values for open space and leisure still existed post-Patrick Abercrombie's County of London Plan (1943). However, Price had a more revolutionary idea of public spaces. Rather than locating a fixed building at the city's core, as Abercrombie proscribed, Price tended towards a temporary solution, for London's evolving neighbourhoods (Shubert, 2005). Building on Price's and Littlewood's concepts of temporality and undetermined programming, the following section discusses the second case study, which focuses on the longer-term temporary architecture of *The Common Room*.

Case study 2: The Common Room

The project *The Common Room* (2014–ongoing), situated in Bow East, London, was initiated as support for a citizen-led high street development. Planned as a long-term project in an area of high land value, public works negotiated a disused, leftover site belonging to Circle Housing Association to be used for a period of three years. It was to provide an open platform for the community to use and set up a public living room as a temporary project. The spatial arrangement of the site was to have a porch, display window/threshold, and a public living room: the public front as a porch became a place for informal engagement, to encounter strangers who may or may not wish to get further involved; the threshold/display window, important as a space where the engagement between the semi-private place of the living room and the public street is negotiated; and the public living room, the most private location of the site at the back and the space for deeper engagement. It was a place where the communities of practice met and where such communities were generated. These took the shape of more formal meetings, organised workshops, debates and platforms for knowledge exchange specific to local needs.

Taking residence and becoming situated in a place within a temporary architecture was the first step in the realisation of *The Common Room*. The project went through a series of naming processes based on the participants involved. In its initial stage, as an extension of a collaborative space between fine art and architecture students at The Cass School of Art, Architecture and Design (London Metropolitan University), the students called it *InterAct Hub*. Once the collaborative project ended after a year it was referred to as the *Roman Road Public Living Room* and then by its current name, given by the local residents, of *The Common Room*. The porch became very much the place where engagement was at its optimum, while the public living room at the back required more committed participants. The public living room was more intimate and those who were happy to be hosted were also happy to engage in the future. The place set for hosting took on two characteristics: firstly, hosting with making, be it food, crafts or poster making, and secondly, as a classroom for learning, be it about oral histories, sanctioned heritage, environment, and debates about placemaking.

The roof and walls that defined and finally created the room made the space drier, yet still unserviced. This meant that staying for any length of time to engage with local residents was hard. No heating, electricity and toilets made the space hard to use for longer engagements.

Before the roof was constructed the space was opened up in the summer as a public space to local citizens. There was Karam (an Asian board game) playing as part of the Roman Road festival, organised by the Roman Road Trust (RRT)[4] residents' group, and the space hosted a series of tea and biscuit mornings and was open for birthday parties. *The Common Room* ran craft workshops, attracting children who brought their parents with them. This became an effective way to engage the parents, especially those from more diverse ethnic backgrounds, to talk about the identity of the town centre. Many of the seemingly middle-class parents commented on the lack of offerings for mothers and children and they showed a strong wish for the workshops to become educational. At one of the sessions there was a desire for crafts to be related to specific local history, environmental knowledge and local needs. When it came to their children, parents had strong opinions and with the rise in young families in the area this became an important consideration for multiple common desires.

A total of four cooking sessions in summer 2016 were open to everyone living and working in and around Roman Road. An invitation was posted on *The Common Room*'s Facebook page (105 followers) and events were tweeted by the RRT to take part in cooking events using waste food from grocery shops on Roman Road. All the products used for the workshops were locally sourced, either donated or salvaged. The workshops attracted a great number of residents and passers-by who wanted to get involved with the weekend food-making programmes. The aim was to expand this to ask residents from a wider area to share family recipes or traditional dishes, which could tell more about their country of origin and their background. *The Common Room* as a classroom hosted children from the adventure playground, it programmed a treasure hunt to explore the area and designed a temporary playground. In collaboration with local residents, public works did a tour of the green spaces behind the high street before coming back to *The Common Room* and undertaking educational workshop on ecology, talking about green spaces as places of environmental learning while making bird feeders. This led to the discussion on the importance of an active use of green spaces for health purposes and biodiversity, leading to a subsequent project with Circle Housing, *The Wilderness project* (2016), with its own independent funding generated by public works through RRT.

A cultural project, *Bow: Her Story* (2016), an initiative by RRT and developed within *The Common Room*, looked at the heritage of women who have made an impact in Bow. Historically, there were many notable women in and around the Roman Road, such as suffragette Sylvia Pankhurst and others in the East London Federation of Suffragettes in 1914, and not forgetting many women engaged in civic activity today (for example, those in the RRT). *Bow: Her Story* has resulted in a project, which at the time of writing is seeking funding to work with multicultural women in Bow. This project aims to use history, hospitality, food, storytelling and performance to engage Somalian, Bengali and Afro-Caribbean women and East End old-age pensioners to discuss their domestic and public life in the UK, hoping to reveal commonalities, conflicts, insecurities, misunderstandings and vulnerabilities. These stories would then be transformed into everyday objects such as wallpaper, tablecloths and a local woman's magazine, whose content and dissemination will be decided and negotiated between the female participants.

A critical awareness of the design and themes of events becomes necessary in order not to alienate different ethnic groups and to allow a critical look at the role of an event in a multicultural neighbourhood. Events are part of the spectacle and narrative of a place and that narrative needs to be designed in a participatory manner. Locations of temporality possess identity, relation and history; as Temel (in Arlt et al., 2006, p.59) says, 'they are not empty; they are screens onto which something is projected'. The question remains: What actions can leave the most traces recognised by its citizens and how do they feel they belong to those traces? For its first year *The Common Room* also accommodated a mobile booth designed by architect Carlotta Novella. It was designed to create a mobile presence on the street, on the one hand, and be a billboard, on the other, to promote the activities of the temporary site. This booth had two functions: firstly, to host residencies for participatory practitioners and researchers, and secondly to disseminate/disclose information/knowledge on its billboard. One aspect that is not discussed with mobile structures is the resources and effort they take to operate and run. Its duration is much more limited and unless considered carefully, it takes more effort to operate than the outputs it generates. Its iconic image (like a circus coming to town) is part of its allure and needs to be carefully considered. In this situation the mobile booth mainly worked as a billboard than a space where residencies can happen. All the events and engagements held at *The Common Room* delivered Fukuyama's (1995, p.10) notion of social capital, 'the ability of people to work together for common purposes in groups and organizations'. The recording of such social capital over time becomes a key tactical factor in transformation of land towards its use as a community asset. Public works also used *The Common Room* as a planning room where local residents mapped the local civic infrastructures, the functions of the shops and the area's insufficient public realm. The project also supported the development of a neighbourhood plan (neighbourhood planning gives communities direct power to develop a shared vision for their neighbourhood and shape the development and growth of their local area), part of its strategic local work.

Conclusion

As mentioned in the introduction, the temporary spatial objects/architectures described in this chapter have tactical ambitions in placemaking. They offer a place where social capital (Fukuyama, 1995) is generated in a single space. The tactics are generated through learning platforms, workshops and events, and once its social capital is quantified and documented communities have the power to negotiate land use as community asset. This agency can enable new forms of public space for social interaction to emerge based on evidence generated by the temporary architectures/objects. Tactical modes of operation alone do not yield the scale of change and influence one may need in a neighbourhood. Alternative ways of strategic thinking that is citizen-led can negotiate with institutional powers that govern. Participatory and collaborative planning practices challenge large-scale or masterplanning approaches. Such approaches, as mentioned by Arlt et al. (2006), focus on 'everyday urbanism' reacting to existing local situations.

Such community assets also have the power to fight threats to future erasure from development. In this sense, spatial objects/architectures become disobedient as they start to exert counter power. The historical examples offered in this chapter have foregrounded the problems of control and power and the need for the spatial practitioner to critically understand the power relations that the spatial project creates. Although the Agitprop trains gave rise to developments in art, theatre and consumerist spectacles, they were not making platforms for democratic social, political engagement that ultimately create places. They used mobility and art towards control of public opinion – indeed,*The Fun Palace*'s automated games theory may have had similar quandary if it had been realised. As a conceptual project, *The Fun Palace* intended to provide a democratic and flexible space, whose form and performance was dictated by the public. This is much closer to the ambitions of 'situated' temporary architecture and spatial objects discussed in the contemporary case studies of this chapter, striving to be politically tactical. One hopes to position such projects between the performative spectacle of the Agitprop but with narrative derived from its citizens, not its governments, and designed with the democratic spatial ideologies of *The Fun Palace*. *The Fun Palace*'s ultimate desire was consensual use of space – it was utopian by not considering the complexities of human psychology. It assumed a common mode of behaviour by all.

To summarise, there are six key factors to consider when proposing spatial objects/architectures described in this chapter: the critical understanding of power relations in a place and the role of spatial objects/architectures within it; designing a programme and events with intentionality equipped with situated knowledge; and considering the potential of projects to be constructed in an incremental way. This slow process enables communities to engage and form opinions; the potential of affordable construction to allow continuous change and adaptation; the development and use of tactical knowledge and its application within strategic agendas to negotiate with formal institutions of power; understanding interests and conflicts and learning how to negotiate them both with institutions of power and citizens; and ensuring to have an independent position in terms of financing of the project and the practitioners affiliation with organisations.

Ultimately these structures allow non-corporate public spaces that can empower and open up possibilities for democratic city spaces (Mouffe et al., 2001). They can be places where conflict and difference are negotiated: for example, Mouffe's (ibid., p.123) agonistic public spaces are disappearing as the neoliberal project takes stronger hold over our city spaces: 'you need to have a choice. What I'm arguing is that this form of agonistic public sphere is not something that should be seen as negative or threatening for democracy.' Every temporary project should consider its own role in the specific place they are situated and its role over time through collective discussion with local residents. The spatial object/architecture can evolve, allowing the social interests and concerns to manifest themselves physically. Such processes can be slow and appear conventionally unproductive. The temporary architecture needs to be tactical; a part of the informal city that negotiates the formal economic and political structures at play. It needs to give the image of productivity to survive. In this sense temporary architecture becomes a way towards a tactical placemaking in neighbourhoods. The practitioner of the temporary spatial object/architectures is permanent; they negotiate the site and build the initial infrastructures, physical and operational, as cheaply and flexibly as possible. This creates the civic platform where its sociability can flourish and grow and protects the site from the encroachment of the formal city negotiating its position with the formal. The role of permanent practitioner and their temporary structures are in constant evolution situated in a place and context.

Notes

1   public works, www.publicworksgroup.net
2   *DIY Regeneration*, www.diyregeneration.net/
3   Grizedale Arts, www.grizedale.org
4   Roman Road Trust, http://romanroadtrust.co.uk/

References

Arlt, P., Haydn, F., and Temel, R. (2006). *Temporary Urban Spaces*. Basel: Birkhäuser.
Bishop, C. (2005). 'Art of the Encounter: Antagonism and Relational Aesthetics', *Circa*, 114.
Bourriaud, N. (2010). *Relational Aesthetics*. Dijon: Les Presses du Reel.
Brown, K. (2013). 'Agitprop in Soviet Russia', *Constructing the Past*, 14(1). Available at: http://digitalcommons.iwu.edu/constructing/vol14/iss1/4. [Accessed: 23 January 2018].
Chomsky, N. and Herman, E. (1988). *Manufacturing Consent*. New York: Pantheon Books.
De Certeau, M. (2011). *Practice of Everyday Life*. Berkeley: University of California Press.
Debord, G. (1970). *The Society of the Spectacle*. Detroit: Black & Red.
Driver, S. and Martell, L. (1997). 'New Labour's Communitarianisms', *Critical Social Policy*, 17(52).
Flood, C. and Grindon, G. (2017). *Disobedient Object*. London: Victoria and Albert Museum.
Fukuyama, F. (1995). *Trust: The Social Virtues and the Creation of Prosperity*. London: Hamish Hamilton.
Latour, B. (2007). *Reassembling the Social*. Oxford: Oxford University Press.
Lave, J. and Wenger, E. (n.d.). *Situated Learning*.
Mathews, S. (2005). 'The Fun Palace: Cedric Price's Experiment in Architecture and Technology', *Technoetic Arts*, 3(2).
Mouffe, C. (1992). 'Citizenship and Political Identity', *October*, 61, 28.

Mouffe, C., Deutsche, R., Joseph, B., and Keenan, T. (2001). 'Every Form of Art Has a Political Dimension', *Grey Room*, 2.

Pipes, R. (1996). *A Concise History of the Russian Revolution*. New York: Vintage Books.

Shubert, H. (2005). 'Cedric Price's Fun Palace as Public Space.' Available at: http:// howardshubert.com/Architecture_Curator/Cedric_Price_files/Cedric Price Fun Palace as Public Space.pdf [Accessed: 30 May 2018].

Spiegel, A. (2015). 'Study Shows that People are Getting Even More Obsessed with Pop-up Restaurant,' *Huffington* Post, 10 April. Available at: www.huffingtonpost.com/2015/04/10/ pop-up-restaurants-study- people-more-obsessed_n_7035394.html [Accessed: 30 May 2018].

Susen, S. (2011). 'Critical notes on Habermas's theory of the public sphere', *Sociological Analysis*, 5(1).

10 Place guarding

Activist art against gentrification

Stephen Pritchard

Abstract

The US policy platform of creative placemaking (Gadwa Nicodemus, 2013, p.214) is becoming a dominant global vehicle for the implementation of neoliberal ideologies (Wilbur, 2015, pp.96-7) like the Creative City (Landry and Bianchini, 1995) and Creative Class (Florida, 2002). This chapter explores how activist art within the context of housing protests can offer potentially generative approaches to place guarding capable of resisting appropriation by policy platforms like creative placemaking.

This chapter argues that creative placemaking is a state- and local-authority inspired policy that is wedded, via corporate partnerships, to neoliberalism: an approach that merges art with community and economic development at every level of society, from the global to the hyper-local. It suggests that creative placemaking thereby utilises Creative City and Creative Class models alongside New Urbanist principles and social capital theory to become an effective means of gentrification. Yet the chapter also argues that art and artists can use their creativity as part of broader social movements to resist and oppose gentrification. This process can be described as a form of 'place guarding', collective acts of protecting existing people and places from the ravages of neoliberalism and policies and practices such as creative placemaking and 'artwashing', the use of art as a veneer or mask for corporate or state agendas (O'Sullivan, 2014).

Introduction

Art has been used as a symbol of property, wealth and power since the beginnings of civilisation (Groys, 2008). However, it became wedded to capitalism and thereby to 'regeneration' agendas during the late twentieth century (Vickery, 2007), culminating in the dominant ideologies of the Creative City (Landry and Bianchini, 1995) and Creative Class (Florida, 2002). Creative placemaking is founded on these notions (Markusen and Gadwa Nicodemus, 2010b, pp.5, 31). Yet the Creative City and Creative Class models are widely accused of driving gentrification globally (Malanga, 2004, p.36; Peck, 2005, pp.740–1). These models have been widely adopted by national and local governments who have corralled the arts into a wildly divergent economic classification commonly known as the creative industries. Indeed, even the originator of the Creative Class recently acknowledged that it led to gentrification (Florida, 2017, pp.xvi–xvii). The co-option of art as one element of the creative industries leaves it vulnerable to economic exploitation, including its use as a placemaking tool for state-led gentrification.

This chapter asserts that gentrification remakes places for the middle-classes and that that art is its stalking horse. Art, when fused with the overlapping agendas of urban renewal, localism and neoliberalism and the normative ideals of the civic, community development, social capital and, ultimately, financial capital, becomes a powerful weapon for displacement. Housing is the key battleground. But some artists, working indistinguishably with other community members, employ acts of resistance to deny compliance, to refuse complicity. Invoking and embodying a spirit of disobedience, indignation and insurrection, they stand in support of those threatened by neoliberalism's insatiable desire for accumulation by dispossession. Such acts can be considered as attempts to defend and guard people and places threatened by gentrification; actions aimed at exposing the complex interrelationships that occur when art and urban regeneration meet in contested spaces.

The roots of creative placemaking

It is possible to argue that creative placemaking is a product of Creative City participation and investment, New Urbanism's 'community building' of diverse neighbourhoods and 'human-scaled urban design' (Congress for New Urbanism, 2017), and Creative Class economics. Moss (2012), for example, argued that Florida and his associates had, in *The Rise of the Creative* Class (2002), not only sold the idea of Creative Class as a means of improving economic prosperity but also 'made the current creative placemaking movement possible.' The creative placemaking 'movement' is particularly strong in the US and expanding globally. The US independent federal arts funding agency, the National Endowment for the Arts, describes creative placemaking as a strategic partnership that uses art and culture to animate and rejuvenate people and places, as well as improve 'local business viability and public safety' (Gadwa Nicodemus, 2012). This is a definite and important shift in tone: 'regeneration' renamed as 'placemaking'; 'cultural regeneration' becomes 'creative placemaking'; and art is reimagined as 'one of the leading place-making devices' in this new urban economy (Mathews, 2010, p.667).

For Courage (2017), placemaking as a form of social practice – as a form of socially engaged art practice – is distinct from creative placemaking because it reflects a recognition of the value of how some artists have worked with people and place for many years (Schumerth, 2015). But, just as there is nothing new about artists exploring place, there is also little new about other sectors using artistic practices for their own benefits – community organisers and property developers alike. Yet whilst Courage acknowledges the value of rigorous, long-standing, socially embedded artistic practice in which artists and community members become (relatively) equal contributors in social practice placemaking processes, the practice has been reduced and rebranded; conflated within creative placemaking. It is possible to argue that the rebranding and appropriation of socially engaged art practice began when Landesman (2009) coined the term 'creative placemaking' to describe a strategic policy platform that McKeown (2016) stated could be considered as part of the ongoing evolving efforts to embed arts-led processes within cultural regeneration initiatives. In this way, creative placemaking becomes a component of neoliberal post-welfare govern-ance; particularly part of localism agendas. It is therefore possible to argue that creative placemaking, like localism more broadly, seeks to consolidate 'the social' within localised free market economics (Hess, 2009; Davoudi and Madanipour, 2013; Williams et al., 2014).

Creative placemaking, social capital and gentrification

Creative placemaking, like the arts in general, promotes itself in dual economic terms: as capable of generating financial capital and social capital. For example, Stern (2014, p.86) claims that within the practice 'there is room for both investment- and social capital-driven policy making', although warns that crea-tive placemakers must fully understand the practice better to avoid unanticipated or unwanted outcomes. Bennett (2014, pp.77–8) adopts a more positivist posi-tion, arguing that creative placemaking can 'strengthen economic development, encourage civic engagement, build resiliency, and/or contribute to quality of life'. Social capital theory is an economic and cultural approach that seeks to build networks and civic norms by developing 'the links, shared values and under-standings in society that enable individuals and groups to trust each other and so work together' (Keeley, 2007, p.102). And it is social capital that is increasingly emphasised as a way of deflecting attention from creative placemaking's primar-ily economic function, whilst simultaneously building its claim to be a tool for community development and civic improvement.

Social capital is considered by the World Bank to be a crucial element of community-driven development, alongside participation and the decentralisation of many state functions to regional and local authorities, including civic society and private sector organisations:

> Social capital refers to the institutions, relationships, and norms that shape the quality and quantity of a society's social interactions. Increasing evi-dence shows that social cohesion is critical for societies to prosper econom-ically and for development to be sustainable. Social capital is not just the sum of the institutions, which underpin a society – it is the glue that holds them together.
>
> (The World Bank, 2004, n.p.)

Artists engaged in community contexts are perfectly suited to harnessing social capital because, unlike corporate consultants, they are frequently able to earn the trust of local people and community groups, and trust is perhaps the single most important element of social capital (Pritchard, 2017). Because social capital theory underpins neoliberal state policies around urban regeneration, community development, civic engagement and localism (Mayer, 2003, pp.119–22) and is also a global economic theory that ultimately seeks to monetise intangible assets and exploit them, the marriage of art and social capital can be considered ideological.

Nevertheless, it is important to understand that social capital is used by the state as a way of measuring public 'good' and public 'bad' (Pantoja, 2002, p.138). It thereby functions positively and negatively. Social housing has suffered from intensive under- and dis-investment for many years and many of these places, now deemed to possess low levels of social capital, are earmarked for improvements, which include arts engagement and wellbeing programmes (Vella-Burrows et al., 2014; Robinson, 2016; Ecorys, 2017). If programmes such as these are, as is almost always the case, unsuccessful, communities may be classified as 'failing' or 'deprived' or 'sink estates' or 'slums' and therefore in need of further social and physical regeneration. Many of so-called 'sink estates' have witnessed a raft of arts projects. In London, for example, the Aylesbury Estate has witnessed numerous participatory art projects supported by the Creation Trust.[1] Furthermore, the Lansbury Estate has similarly been the site of arts interventions, including a long-term Victoria and Albert Museum (V&A) 'micro-museum', whilst the V&A residency at Robin Hood Gardens culminated with the removal of a section of ex-council housing to be preserved and rehoused in its new museum in Stratford, East London.

Within a neoliberal agenda, the standard approach to improve a neighbourhood's social capital is to renew it by bringing in new, middle-class people with high social capital (Josten, 2013, p.21). This is fertile ground for artists and creative placemakers. But the presence of artists and placemaking initiatives often drives gentrification and the area's original residents are often ultimately displaced (Bedoya, 2013). In these cases, artists and creative placemakers can perform a dual role in what can be considered as artwashing: first, they 'embed' within local communities and carefully and creatively harvest their social capital; second, their presence acts as an indicator of the area's creativity and a flag to other property developers and small enterprises (often established by those classed as 'hipsters') that the area is 'up-and-coming' (Pritchard, 2017). It is this twin-pronged manipulation of artistic practices that are increasingly opposed and called out by activists.

Cultural activists and activist artists understand that artistic practices like socially engaged art and creative placemaking encourage artists and arts organisations to work in partnership with private-, public- and third-sector institutions (Balfron Social Club, 2015). This leads commissioned artists to become complicit in capitalist exploitation, reproducing the state and corporate agendas and promoting forms of neoliberal governance that are at once more central and more local. The now ubiquitous Creative City has become an emancipatory ideal for the new Creative Class and a revanchist reality for the city's 'others' (Lees, 2014, p.52), with artists often touted as exemplars of a new class of creative entrepreneurs. This class polarisation, driven by a global free market economy, restricts community organisations (including arts organisations) tasked with tackling exclusion and poverty to the limited objectives set by the state and corporations (Mayer, 2007, p.100). Peck (2009 [2007], pp.5–7) describes this as 'the creativity script' – individualistic and market-driven – in which creativity becomes a cheap and cheerful means of reinforcing fierce competition, valorising a 'superior' Creative Class, and validating it via the market and 'post-progressive urban policy'. This incorporation of art within the discourse of neoliberal urban development (and therefore gentrification) makes it increasingly complicit with state and corporate interests. The narrative is one of 'heroic elites' saving inner cities from the 'dangerous classes' (Madden, 2013): the binary opposite of Lefebvre's (1968) assertion of the right to the city.

There are deeply problematic links between creative placemaking and third-wave gentrification. For Smith, the third wave – or 'generalised gentrification' – expands global capital's reach into local neighbourhoods using complex amalgamations of 'corporate and state powers and practices' to create 'landscape complexes' that integrate 'housing with shopping, restaurants, cultural facilities … open space, employment opportunities' and more. Generalised gentrification enables 'global interurban competition' couched in the language of 'urban regeneration' (Smith, 2002, pp.441–3). Gentrification foreshadows the 'class conquest of the city' with 'new urban pioneers' intent on the 'systematic eviction' of the working class from cities. Social histories are rewritten 'as a pre-emptive [sic] justification for a new urban future'; slums transform into 'historic brownstones' with façades 'sand-blasted to reveal a future past' as the middle classes 'recolonize the city' (ibid., 1996, p.25–7).

Creative placemaking as neoliberal function

The neoliberal language of urban regeneration

Creative placemaking merges the discourse of urban regeneration with the language of urban architects, business and economics as well as that of not-for-profit community development, softening neoliberalism with culture and community on the one hand, and neoliberalising culture and community on the other. Typically, creative placemaking advocates a 'more decentralized portfolio of spaces acting as creative crucibles' (rather than the traditional big arts centre or cultural quarter models) in which arts and culture exists 'cheek-by jowl with private sector export and retail businesses and mixed-income housing, often occupying buildings and lots that had been vacant and under-used' (Markusen and Gadwa Nicodemus, 2010b, p.3). But these sorts of 'pop-up' art activities and forms of mixed-use housing and community development are often associated with gentrification (Hancox, 2014; Schulkind, 2017). Creative placemaking in this vein thereby reproduces a falsely utopian vision in which artists live next door to corporations, rich next to poor.

Referring to fourteen specific case studies, in their creative placemaking White Paper, Markusen and Gadwa Nicodemus stated: 'Private sector developers, lenders, sponsors, philanthropists, and local arts businesses have in most cases been important facilitators of arts and culture-led revitalization' (2010a, p.20). The voices of community members, individual artist practitioners and cultural activists are obviously missing from this description, revealing the top-down nature of placemaking (and cultural regeneration in general). The link to property developers is telling. Property developers and local authorities often utilise the 'artistic mode of production' as symbolic capital with which to construct 'new place-identities' that increase their economic value by rebranding them as 'creative' (Zukin and Braslow, 2011, p.131). Typically, gentrification and displacement follow as can be seen, for example, in the rebranding of Manhattan's SoHo district, London's Kings Cross as the city's 'Knowledge Quarter',[2] Chicago's Wicker Park, and Queen Elizabeth Olympic Park, London.[3]

Community engagement and social capital

Creative placemaking often markets itself as employing a 'community arts' approach that can, for example, 'impact participants individually by fostering transferable life skills' and 'cultivate social capital' (Artscape, 2017). This is somewhat of a departure from traditional notions of community arts practice, perhaps reflecting what Kelly considered to be the demise of community arts as activism, replaced by a professionalised and respectable simulacrum (1984, p.38). However, Artscape's normative outlook, which is mirrored by some other creative placemaking agencies, not only appropriates the contested term 'community arts', it does so to mask its (frequently downplayed) economic objectives. Some placemakers are, however, rightly sceptical about privileging artists by placing them at the centre of placemaking (Nikitin, 2013; Project for Public Spaces, 2012). Nonetheless, both movements tend to emphasise the importance of social capital, repeating arguments about how social capital is a 'social good' capable of combating social exclusion by increasing participation and citizenship. Project for Public Spaces (PPS) (2012), for example, argues that 'one of the most important factors in any effort to change the way that we shape the places where we live and work … [is] social capital.' Placemaking becomes a way to develop 'loose social networks' and thereby 'urban resilience' and to reinforce existing social ties whilst simultaneously stimulating new ones (ibid.). The importance of social capital developed through community participation to placemaking cannot be underestimated.

Mayer describes the rush by policymakers to broadly apply the rhetoric of grassroots participation and community activism as part of a 'new mode of governance' dominated by economics:

> By prioritizing specific forms of civic engagement (while neglecting others), [social capital] filters the contemporary reconfigurations in the relationship of civil society, state and market in a peculiar way, which is conducive to

supporting the spread of market forces in areas so far beyond the reach of capital.

(Mayer, 2003, pp.110-1)

Insisting that marginalised people and communities suffer from a lack of social capital rather than disempowerment, domination and exploitation enables neoliberal governments to employ the burgeoning third sector to encourage wealthier people to volunteer. This undermines direct state welfare systems, expanding the reach of the market way beyond its traditional confines. Social capital policy achieves this by labelling voluntary and civic activities as a form of capital, which reconfigures them as economic assets, thereby facilitating 'good governance' based upon public-private-civic partnerships (Mayer, 2003). Social capital therefore offers low-cost solutions for social issues, placing social capitalists at the head of efforts to both reduce the size of the state and to refine traditional neoliberalism. Social capital interventionism, spearheaded by state-led arts institutions and creative placemakers, forms a vanguard of (relatively) cheap, often cheerful options that not only 'engage' (some) marginalised people but also produce aesthetically pleasing outcomes which in turn are used as 'evidence' of 'inclusion'.

This increasing valorisation of civic engagement embeds the principle of social capital within arts and community organisations as a universally positive output. Social capital serves to distract from the economic and political processes that drive the civic engagement agenda. Refocusing attention onto marginalised and excluded individuals rather than the causes of their marginalisation and exclusion, it recruits disadvantaged people as the 'agents' of their own salvation – social capitalists 'whose "belonging" is conditional on their mobilizing the only resources they have as a form of capital'. This is because accumulating social capital is not intended to benefit the poor or working-classes. The goal of social capital is about '"empowerment" and "inclusion"' rather than 'economic security for the poor or the reduction of inequality' (Mayer, 2003, p.123). The democratic façade of social capital policy offers the perfect means of implementing other neoliberal agendas such as austerity, devolution and localism. It is therefore important to understand that any attempt to incorporate social capital policy within creative placemaking can only function to reinforce the neoliberal project and enforce neoliberalism upon the poorest and most vulnerable in society.

In this way, austerity offers many 'opportunities' for arts organisations and artists to exploit communities using a façade of 'community benefit' and 'social impact'. In this sense, creative placemaking can be thought of as exploiting the intangible assets in a neighbourhood faced with or in the early stages of gentrification; creative placemakers become its agents, bringing about a form of social change that is antithetical to the principles of social justice. Using processes dressed as 'community-led development', creative placemaking encourages public/private initiatives to integrate art, creative consultation, urban design, participatory democracy, civic governance and economic revitalisation. Creativity becomes part of the neoliberal 'market-building project' that rolls-back some social and institutional functions and rolls-out creative- and community-led alternatives as part of its 'destructively creative social order' (Peck and Tickell, 2007, pp.33–4). There is, however, a vast chasm between, for example, social capital, community governance and creative consultation and the radical demands for social justice, to take back the city and for the right to the city.

Artists against gentrification

Creative placemaking is ultimately a state- and local authority-led policy, filtered down to arts institutions and then on to artists and down to participants and audiences via agencies, funders and a raft of philanthropic foundations and charities. Agencies such as Artscape develop projects that typically seek to 'empower' marginalised people and places through a mixture of socially engaged art, education and outreach (Artscape, 2017). They are often implemented in areas classified as in need of, or already undergoing, regeneration initiatives, and art has long been perceived as a primary tool for urban regeneration. From another perspective, this could equally be seen as targeting low- or under-engaged people and communities deemed to be lacking in culture (and therefore social capital). Thus, artists and arts organisations are funnelled by funders and agencies into working in places identified by government, with people categorised as in need by government, addressing problems defined by government, using approaches and outcomes prescribed by government. Art and artists become instrumentalised by these processes.

To combat this, community activists – often artists who deny their status as artists who work collectively using pseudonyms – employ creative techniques of resistance such as tactical media and anti-art strategies to resist the traditional agents of gentrification. Their creative resistance extends beyond the traditional drivers of gentrification (the state, local authorities and corporations). They oppose those who often operate as socially engaged artists or creative placemakers in the service of gentrifiers – whether local 'micro-enterprises' or large-scale state and corporate 'regeneration' initiatives. This is because cultural activists recognise that art is increasingly used to smooth and gloss over social cleansing and gentrification, functioning as 'social licence' (Evans, 2015, pp.70–84), a public relations tool and a means of pacifying local communities: a practice known by activists as 'artwashing' (Pritchard, 2017).

Rather than becoming complicit in neoliberalism, artists and communities must, as Harvey (2008, p.23) proposed, 'exercise ... collective power over the processes of urbanization' and assert our right 'to make and remake ourselves and our cities'. This is because our freedom to be able to determine how we live in our cities, our places, our spaces is 'one of the most precious yet most neglected of our human rights'. Reclaiming 'the right to the city' is a radical political act (Lefebvre, 1968, p.158). It requires self-organisation, self-realisation, self-determination. It requires social practice as activism – 'the practice of everyday life' (De Certeau, 1984). This practice must, as Holmes (2012, p.79) proposes, produce a common desire to change the way we live 'without any guarantees'. We must reclaim our freedom to make and remake our spaces from those who wish to define and limit our everyday lives through a totalising system of 'controlled consumption' and 'terror-enforced passivity' (Lefebvre, 2000, pp.196–7). For many, contemporary social spaces enforce what Lefebvre describes as the 'latent irrationality beneath an apparent rationality, incoherence beneath an ideology of coherence, and sub-systems or disconnected territories linked together only by speech' (ibid., p.197). We must therefore understand the complex dialectics of neoliberal space.

Creative placemaking does not and cannot offer people the freedom to take back the city because it is often rolled out as an integral part of neoliberalism's totalising system. It is about 'economic development, livability, and cultural industry competitiveness' (Markusen and Gadwa, 2010b, p.6). Creative placemaking uses art to window-dress neoliberal regeneration and renewal agendas. Activist art, on the other hand, is about direct action: 'full spectrum resistance' (Verson, 2007, p.171). For Verson, cultural activism is a 'living practice' that 'addresses complicated questions about how we build the world that we want to live in' (ibid., p.174). It requires what Jordan and Frémeaux (in Hopkins, 2015) describe as an 'insurrectionary imagination' that recontextualises our understanding of past political movements through a constant striving for the reconnection of art and protest, rather than their artificial separation. This form of activist art pursues social justice and radical political change. It leads activist artists to ask: Why should we (re)make your places for you? People, communities and cultures already exist in places. They produce and reproduce their own social spaces. So why do we need creative placemakers? And if creative placemaking is irrevocably linked to gentrification and social cleansing – to artwashing – shouldn't it be opposed as another Trojan Horse for dispossession?

It is, however, difficult for artists to create anti-gentrification interventions that can avoid recuperation by both an insatiably voracious art world and an exploitative capitalist machine. Nevertheless, there are many examples of practices that have avoided recuperation and, on some occasions, successfully resisted gentrification. For example, activist art collective BAVO's project *Plea for an unCreative City* (2016) attacked Rotterdam's decision to follow, like Hamburg, a Creative City model which valorised Florida's concept of a Creative Class. US-based activist art collective Illuminator 99[4] often targets gentrification and other issues of social justice using a high-powered, van-mounted digital projector. Its actions are often performed with local people. Temporal employment of symbols, statements and political slogans makes it difficult for these interventions to be recuperated and, indeed, classed as art. Meanwhile, calls for social justice are more accessible to people outside the art world; to local people threatened by displacement and inequity. The next section outlines three examples of collective interventions involving artists, which could be considered as place guarding.

Balfron Social Club

Balfron Social Club (BSC)[5] is a collective of artists and local people who campaign against the social cleansing of Poplar in Tower Hamlets, London. Its demand for 50 per cent social housing in Ernő Goldfinger's Brutalist ex-council housing icon, the now gentrified Balfron Tower, challenged developers and raised awareness but ultimately failed to prevent the building from becoming luxury housing. Nevertheless, its sharp critique of the role of socially engaged artists as 'placemakers' led to what is a crucial article on the role of socially engaged artists and placemakers and their involvement in artwashing: *Brutalism [redacted] – Social Art Practice and You* (Balfron Social Club, 2015). For BSC, artists today are faced with a choice: 'join the club to make ends meet' or become 'completely marginalized and unable to work'. The collective is here referring to the immense pressure many artists feel to conform to policies such as creative placemaking and arts-led regeneration agendas to make a living.

The 'perfect storm' of austerity, increased corporate interest and the professionalisation of the arts as the 'creative industries' led, according to BSC, to 'the birth of the community based Social Art Practitioner' (2015). Funded by the state and employed by local authorities and property developers, these artists are placed into communities facing gentrification to operate as the foot soldiers of social cleansing (ibid.). The collective's position offers a completely alternative view of a practice – socially engaged art – that is lauded internationally as an almost universally positive act of change or placemaking. For BSC, the process of creatively engaging citizens, stimulating employability and building 'positive social change' parrots those of housing developers, housing associations, councils and arts funders. Interestingly, BSC links social capital to enterprise and placemaking policy: a top-down strategy presented, through the filter of creative engagement, as 'grassroots' (ibid.). BSC has raised and continues to raise the issue of the sell-off of social housing to private luxury housing developers and the politics that underwrite this situation. Their work at Balfron Tower, across Tower Hamlets and beyond reveals how artists with other community members attempt to guard their homes and communities against the onslaught of state-led gentrification.

Southwark Notes Archive Group

Southwark Notes Archive Group (SNAG)[6] is a politically independent group of 'local people who aren't particularly happy about what is going on in the name of "regeneration"' (Southwark Notes, 2017). It demands that regeneration schemes enhance the quality of existing communities and opposes gentrification. Like BSC, the collective uses a moniker to maintain anonymity. SNAG is careful to avoid 'political and intellectual language' in its campaigning, thereby ensuring it is as accessible to as many people as possible. The group works to oppose gentrification across Southwark and further afield, both individually and with other groups. It also has links to anti-gentrification movements in other countries, including international activist art collective Ultra-red.[7] SNAG's opposition to the 'redevelopment' of the Heygate Estate – a large ex-council housing estate in Walworth, South London – was extensive. It occupied the estate's gardens, organised regular walks around the area, helped community groups fighting the demolition of the nearby Aylesbury Estate and played an important role in the successful scrapping of artist Mike Nelson's planned pyramid sculpture which was to have been constructed from the detritus of a demolished building on the estate. Whilst not all the collective are artists, SNAG collaborates with residents in campaigns, sharing cultural awareness and labour, facilities and ways of resisting. This mode of working means artists become virtually indistinct from other activists (Graham and Vass, 2014, p.16). This is a form of resistance that seeks to collapse art into everyday life.

Boyle heights alliance against artwashing and displacement

The most prominent example of anti-gentrification and anti-artwashing campaigning is Boyle Heights Alliance Against Artwashing and Displacement (BHAAAD).[8] BHAAAD vigorously campaigns for all art galleries in Boyle Heights, Los Angeles, to leave, accusing them of artwashing social cleansing (Aron, 2016; Barragan, 2016; Stromberg, 2016). The alliance comprises several organisations: Union de Vecinos; Defend Boyle Heights; The Eastside Local of the Los Angeles Tenants Union; School of Echoes Los Angeles; and what BHAAAD describes as 'Multiple Affinity Groups of Artists'. These artists acknowledge the role art plays in gentrification and refuse to be used to artwash 'the realities of racial and economic violence' (Boyle Heights Alliance Against Artwashing and Displacement, 2017). BHAAAD has raised the issue of the struggle of residents to guard their neighbourhoods against an initial influx of arts organisations and artists that brings further gentrification in its wake. Some long-standing arts organisations are employing creative placemaking in attempts to preserve the local, predominantly Latino culture in Boyle Heights (Jackson, 2015), but they too are finding themselves targets for BHAAAD. The constantly shifting spaces and relationships in which anti-gentrification activism and place guarding take place are complex and deeply contested. The function of artwashing and the instrumentalisation of art in the service of regeneration in all its forms serve to fragment artists and communities.

Demanding the impossible

There are many other examples of radical art/cultural activism that utilise similar modes of direct action against gentrifiers and placemakers in attempts to guard complex community structures and rights and protect existing ways of living. They share a common belief that it is time for 'the dispossessed to take back control of the city from which they have for so long been excluded' (Harvey, 2010, p.32). To some, this may seem utopian; to others, 'Demanding the impossible may be … as realistic as it is necessary' (Pinder, 2015 [2013], p.43). The 'urban renaissance' narrative is not only insidious but, as Madden (2013) argues, 'a condescending and often racist fantasy'.

Conclusion

This chapter has argued that creative placemaking is a neoliberal vehicle for accumulation by dispossession and gentrification. It has traced a path from art as status symbol – an object of property and power – to art as a process that transfers tangible assets such as property and places and intangible assets such as the social capital of communities into the hands of the powerful and wealthy. It has argued that creative placemaking does not demand the right to the city but rather encourages us to gift our cities to the Creative Class so that they can (re) make places for middle-class gentrifiers. Creative placemaking thereby becomes a policy and practice increasingly adopted and co-opted by state, local authorities and even property developers that use art, design, marketing and community engagement as a way of disempowering people by imposing subtle forms of compliance and implicating communities by securing their participation in the creative destruction of their own neighbourhoods and the dispossession of their own homes.

In this sense, creative placemaking is urban policy in the guise of a gilded neoliberal Trojan Horse. It ushers in the strategic reshaping of the social, spatial and economic characters of neighbourhoods with the sweet cupcakes and pretty bunting of art, the mixed economy and mixed-use, walkable, bikeable spaces. Socially engaged artists are its foot soldiers and amateur artists and participants its conscripts. This chapter has argued that creative placemaking is creative compliance. And compliance kills creativity because, as Winnicott (1991 [1971], p.65) explained, compliance is the opposite of living creatively – 'a sick basis for life'. It is the very real threat of dispossession and displacement, which drives activist artists and cultural activists, like those briefly discussed here, to use anti-art tactics to combat the co-option of creativity by neoliberalism. The radical yet everyday approaches of groups such as BSC and SNAG to working collectively with community members represent attempts to build relationships and guard places from the forces of capital. It is possible to conceive of such actions as 'place guarding'.

Conversely, this chapter makes the case for artistic acts of resistance that deny compliance, invoking and embodying a spirit of disobedience, indignation and insurrection. Such creative acts of refusal can create what might be understood as 'potential spaces' in which radically alternative ways of being and living can develop democratically. Yet activist artists/cultural activists are aware that they cannot effectively challenge an authoritarian neoliberal state or global corporate interest unless they can become part of a much broader social movement. Nevertheless, their tactics of constantly harrying and undermining the power of the state and gentrifiers can win small victories, garnering significant press and political attention and helping to secure concessions for local people embroiled in gentrification and threatened by displacement. These artists are, however, also becoming increasingly aware that their activism can be appropriated or recuperated by the state and corporate interests; disavowing the art in their work to subvert this process.

We must therefore acknowledge that the right to the city is 'an empty signifier' that can be claimed by 'financers and developers' but, equally, by 'the homeless and the *sans papiers*' (Harvey, 2012, p.xv). And, true to their roots, some artists stand in support of those threatened with rehousing; demanding that the dispossessed take back the city. They stand against vested interests, taking direct action with people against placemakers, guarding complex community cultures and their existing ways of living. It is time to demand and to action 'place guarding' rather than 'placemaking' or 'place keeping' (creative or otherwise); to directly resist the neoliberal machinations of gentrification instead of following a hopefully misguided urban (re)map – a simulacrum in which, I have argued, all roads lead to capitalist complicity.

Notes

1 Creation Trust, www.creationtrust.org
2 Kings Cross, London, www.kingscross.co.uk
3 Queen Elizabeth Olympic Park, London, www.queenelizabetholympicpark.co.uk
4 Illuminator99, http://theilluminator.org/about/
5 Balfron Social Club, http://balfronsocialclub.org
6 Southwark Notes Archive Group, https://southwarknotes.wordpress.com
7 Ultra-red, www.ultrared.org/mission.html
8 is Boyle Heights Alliance Against Artwashing and Displacement, http://alianzacon traartwashing.org/en/bhaaad/

References

Aron, H., 2016. *Boyle Heights Activists Demand That All Art Galleries Get the Hell Out of Their Neighborhood.* [Online] Available at: www.laweekly.com/news/boyle-heights-acti vists-demand-that-all-art-galleries-get-the-hell-out-of-their-neighborhood-7134859 [Accessed: 11 March 2017].
Artscape, 2017. *Approaches to Creative Placemaking.* [Online] Available at: www.artscape diy.org/Creative-Placemaking/Approaches-to-Creative-Placemaking.aspx [Accessed: 9 February 2017].
Balfron Social Club, 2015. *Brutalism [redacted] – Social Art Practice and You.* [Online] Available at: http://50percentbalfron.tumblr.com/post/116281372004/brutalism-redacted-social-art-practice-and-you [Accessed: 13 April 2015].
Barragan, B., 2016. *Boyle Heights Activists Want to Banish All Art Galleries.* [Online] Available at: http://la.curbed.com/2016/7/14/12191266/boyle-heights-art-galleries-gentri fication[Accessed: 11 March 2017].
BAVO, 2006. *Plea for an uncreative city.* [Online] Available at: www.bavo.biz/texts/view/ 156[Accessed: 6 February 2016].
Bedoya, R., 2013. 'Placemaking and the politics of belonging and dis-belonging', *GIA Reader*, 24(1).
Bennett, J., 2014. 'Creative placemaking in community planning and development: An introduction to Artplace America', *Community Investment Development Review*, 10(2).
Boyle Heights Alliance Against Artwashing and Displacement, 2017. *Boyle Heights Alliance Against Artwashing and Displacement: The Short History of a Long Struggle.* [Online] Available at: http://alianzacontraartwashing.org/en/coalition-statements/bhaaad-the-short-history-of-a-long-struggle [Accessed: 11 March 2017].
Congress for New Urbanism, 2017. *What is New Urbanism?* [Online] Available at: www. cnu.org/resources/what-new-urbanism [Accessed: 22 March 2017].
Courage, C., 2017. *Arts in Place: The Arts, the Urban and Social Practice.* London: Routledge.

Davoudi, S. and Madanipour, A., 2013. 'Localism and neo-liberal governmentality', *Town Planning Review*, 84(5).

De Certeau, M., 1984. *The Practice of Everyday Life*. Berkeley: University of California Press.

Ecorys, 2017. *Creative People and Places End of Year 3 Evaluation Report: Impact, Outcomes and the Future at the End of Year 3*, London: Arts Council England.

Evans, M., 2015. *Artwash: Big Oil and the Arts*. London: Pluto Press.

Florida, R., 2002. *The Rise of the Creative Class – And How It Is Transforming Leisure, Community and Everyday Life*. New York: Basic Books.

Florida, R., 2017. *The New Urban Crisis: How Our Cities are Increasing Inequality, Deepening Segregation, and Failing the Middle Class - and What We Can Do about It*. New York: Basic Books.

Focus E15, 2017. *Focus E15: Social Housing not Social Cleansing*. [Online] Available at: http://focuse15.org [Accessed: 26 April 2017].

Gadwa Nicodemus, A., 2012. *Creative Placemaking 2.0*. [Online] Available at: www.giarts. org/article/creative-placemaking-20 [Accessed: 2 April 2014].

Gadwa Nicodemus, A., 2013. 'Fuzzy vibrancy: Creative placemaking as ascendant US cultural policy', *Cultural Trends*, 22(3–4).

Graham, J. and Vass, N., 2014. 'Intervention/Art P|Art|Icipate - Kultur aktiv gestalten', 5.

Groys, B., 2008. *Art Power*. Cambridge, Massachusetts: The MIT Press.

Hancox, D., 2014. *Fuck Your Pop-Up Shops*. [Online] Available at: www.vice.com/ en_uk/read/shipping-container-elephant-park-dan-hancox [Accessed: 18 September 2016].

Harvey, D., 2008. 'The right to the city', *New Left Review*, September-October, 53.

Harvey, D., 2010. 'The right to the city: From capital surplus to accumulation by dispossession' in Banerjee-Guha, S., ed. *Accumulation by Dispossession: Transformative Cities in the New Global Order*. New Delhi: Sage.

Harvey, D., 2012. *Rebel Cities: From the Right to the City to the Urban Revolution*. London: Verso.

Hess, D. J., 2009. *Localist Movements in a Global Economy: Sustainability, Justice, and Urban Development in the United States*. Cambridge, MA and London: The MIT Press.

Holmes, B., 2012. 'Eventwork: The fourfold matrix of contemporary social movements' in Thompson, N., ed. *Living as Form: Socially Engaged Art from 1991-2011*. Cambridge Massachusetts and London: The MIT Press

Hopkins, R., 2015. *Isabelle Frémeaux, John Jordan and the rise of the insurrectionary imagination*. [Online] Available at: www.transitionnetwork.org/blogs/rob-hopkins/2015-04/isabelle-fr-meaux-john-jordan-and-rise-insurrectionary-imagination [Accessed: 7 October 2015].

Jackson, M. R., 2015. *How Creative Placemaking Plays a Role in the Creative Economy*. [Online] Available at: www.kcet.org/shows/artbound/how-creative-placemaking-plays-a-role-in-the-creative-economy [Accessed: 24 November 2017].

Josten, S. D., 2013. 'Middle-class consensus, social capital and the fundamental causes of economic growth and development', *Journal of Economic Development*, 38(1).

Keeley, B., 2007. *Human Capital: How What You Know Shapes Your Life*. Paris: OECD Publishing.

Kelly, O., 1984. *Community, Art and the State: Storming the Citadels*. London: Comedia.

Landesman, R., 2009. *Keynote Speech GIA Conference, Navigating the Art of Change*. Brooklyn NY, Grantmakers in the Arts.

Landry, C. and Bianchini, F., 1995. *The Creative City.* London: Demos.

Lees, L., 2014. 'The "new" middle class, lifestyle and the "new" gentrified city', in Paddison, R. and McCann, E., eds. *Cities and Social Change: Encounters with Contemporary Urbanism.* London: SAGE.

Lefebvre, H., 1968. *Le Droit à la ville.* Paris: Éditions Anthropos.

Lefebvre, H., 2000. *Everyday Life in the Modern World.* London: Athlone.

Madden, D., 2013. *Gentrification doesn't trickle down to help everyone.* [Online] Available at: www.theguardian.com/commentisfree/2013/oct/10/gentrification-not-urban-renaissance [Accessed: 20 December 2015].

Malanga, S., 2004. 'The curse of the creative class', *City Journal*, Winter.

Markusen, A. and Gadwa, A., 2010a. *Creative Placemaking.* Washington: National Endowment for the Arts.

Markusen, A. and Gadwa, A., 2010b. *Creative Placemaking: Executive Summary.* Washington: National Endowment for the Arts.

Mathews, V., 2010. 'Aestheticizing space: Art, gentrification and the city', *Geography Compass*, 6(4).

Mayer, M., 2003. 'The onward sweep of social capital: Causes and consequences for understanding cities, communities and urban movements', *International Journal of Urban and Regional Research*, 27(1)

Mayer, M., 2007. 'Contesting the neoliberalization of urban governance', in Leitner, H., Peck, J., and Sheppard, E. S., eds. *Contesting Neoliberalism: Urban Frontiers.* New York: The Guildford Press.

McKeown, A., 2016. *Creative Placemaking: How to Embed Arts-Led Processes within Cultural Regeneration?* [Online] Available at: www.seismopolite.com/creative-place making-how-to-embed-arts-led-processes-within-cultural-regeneration[Accessed: 30 December 2017].

Moss, I. D., 2012. *Creative Placemaking Has an Outcomes Problem.* [Online] Available at: http://createquity.com/2012/05/creative-placemaking-has-an-outcomes-problem [Accessed: 30 December 2017].

Nikitin, C., 2013. *All Placemaking Is Creative: How a Shared Focus on Place Builds Vibrant Destinations.* [Online] Available at: www.pps.org/reference/placemaking-as-com munity-creativity-how-a-shared-focus-on-place-builds-vibrant-destinations[Accessed: 14 February 2017].

O'Sullivan, F., 2014. *The Pernicious Realities of 'Artwashing'.* [Online] Available at: www. citylab.com/housing/2014/06/the-pernicious-realities-of-artwashing/373289/[Accessed: 26 September 2016].

Pantoja, E., 2002. 'Qualitative analysis of social capital: The case of community development in coal mining areas in Orissa, India', in Grootaert, C. and Van Bastelaer, T., eds. *Understanding and Measuring Social Capital: A Multidisciplinary Tool for Practitioners.* Washington, DC: The World Bank.

Peck, J., 2005. 'Struggling with the Creative Class', *International Journal of Urban and Regional Research*, 29(4).

Peck, J., 2009 [2007]. 'The creativity fix', *Variant*, Spring, 34.

Peck, J. and Tickell, A., 2007. 'Conceptualizing neoliberalism, thinking Thatcherism', in Leitner, H., Peck, J., and Sheppard, E. S., eds. *Contesting Neoliberalism: Urban Frontiers.* New York: The Guildford Press.

Pinder, D., 2015 [2013]. 'Reconstituting the possible: Lefebvre, utopia and the urban question', *International Journal of Urban and Regional Research*, 39(1).

Pritchard, S., 2017. *Artists Against Artwashing: Anti-Gentrification and the Intangible Rise of the Social Capital Artist*. [Online] Available at: http://colouringinculture.org/blog/ artistsagainstartwashing [Accessed: 7 October 2017].

Project for Public Spaces, 2012. *How "Small Change" Leads to Big Change: Social Capital and Healthy Places*. [Online]. Available at: www.pps.org/blog/how-small-change-leads-to-big-change-social-capital-and-healthy-places [Accessed: 10 February 2017].

Robinson, M., 2016. *Faster, but Slower Slower, but Faster: Creative People and Places Learning 2016*, Stockton-on-Tees: Thinking Practice (for Creative People and Places). [Online]. Available at: www.creativepeopleplaces.org.uk/sites/default/files/Faster%20but %20Slower_0.pdf [Accessed: 27 May 2018].

Schulkind, R., 2017. *Should we Blame Art for Brixton's Gentrification?* [Online] Available at: www.newstatesman.com/politics/economy/2017/07/should-we-blame-art-brixtons-gentrification [Accessed: 30 December 2017].

Schumerth, C., 2015. *Indianapolis workshops combine placemaking, social practice*. [Online] Available at: http://circlespark.org/placemaking/ [Accessed: 9 November 2015].

Smith, N., 1996. *The New Urban Frontier: Gentrification and the Revanchist City*. London: Routledge.

Smith, N., 2002. 'New globalism, new urbanism: Gentrification as global urban strategy', *Antipode*, 34(3).

Southwark Notes, 2017. *Who Are We?* [Online] Available at: https://southwarknotes.word press.com/where-we-are-coming-from[Accessed: 17 January 2017].

Stern, M. J., 2014. *Measuring the Outcomes of Creative Placemaking*. Washington, DC: Goethe Institut and EUNIC.

Stromberg, M., 2016. *Anti-Gentrification Coalition Calls for Galleries to Leave LA's Boyle Heights*. [Online] Available at: http://hyperallergic.com/314086/anti-gentrification-coali tion-calls-for-galleries-to-leave-las-boyle-heights [Accessed: 11 March 2017].

The World Bank, 2004. *Community Driven Development*. [Online]. Available at: www. worldbank.org/en/webarchives/archive?url=httpzzxxweb.worldbank.org/archive/websi te00996A/WEB/OTHER/COMMUNIT.HTMandmdk=21600690 [Accessed: 24 April 2017].

Vella-Burrows, T. et al., 2014. *Cultural Value and Social Capital: Investigating Social Capital, Health and Wellbeing Impacts in Three Coastal Towns Undergoing Culture-Led Regeneration*. Folkestone: Sydney De Haan Research Centre for Arts and Health and Nick Ewbank Associates (for the AHRC Cultural Value Project).

Verson, J., 2007. 'Why we need cultural activism', in Collective, T. T., ed. *Do It Yourself: A Handbook for Changing Our World*. London: Pluto Press.

Vickery, J., 2007. *The Emergence of Culture-Led Regeneration: A Policy Concept and Its Discontents*. Warwick: The University of Warwick.

Wilbur, S., 2015. 'It's about time creative placemaking and performance analytics', *Performance Research*, 20(4).

Williams, A., Goodwin, M., and Cloke, P., 2014. 'Neoliberalism, Big Society, and progressive localism', *Environment and Planning*, 46.

Winnicott, D. W., 1991 [1971]. *Playing and Reality*. London and New York: Tavistock/ Routledge.

Zukin, S. and Braslow, L., 2011. 'The lifecycle of New York's creative districts', *City, Culture and Society*, 2(3).

11 Outros Espaços

Apathy and lack of engagement in participatory processes

Luísa Alpalhão

Abstract

Participation has become an often meaningless buzzword associated with notions of empowerment and democratic values (Blundell-Jones, Petrescu, and Till, 2005, p.xiii). It is recurrently taken for granted that people want to participate in the making of their cities and to be actively involved in shaping their environments (Beebeejaun, 2016, p.7). In this chapter, the position that participation is effectively welcomed and desired by all is explored and reflected upon through the problematics of apathy and lack of engagement found in the participatory project *Outros Espaços* (*Other Spaces*) (2014-15), in Beja, Portugal. *Outros Espaços* was an initiative of Atelier Urban Nomads in collaboration with local authorities, local non-governmental organisations (NGOs), local residents' association, and a local school. It is an illustrative case study where despite purportedly being a collective, collaborative participatory process, the participants were often reluctant to get involved in the transformation of their neighbourhood's public spaces. *Outros Espaços* found that passivity, lack of engagement and even a certain level of apathy can be present in projects that intend to empower the citizens. Collaboration, in this case, did not exist, nor were the neighbourhood's public spaces socially or spatially transformed in a long-lasting way. The meaning of success in this instance is questionable and this chapter presents a critique of the project's process to reveal some of the pitfalls of architectural practices that seek to develop social agendas and asks why this lack of interest, and in some cases apathy, prevailed.

Introduction: Participation, engagement and placemaking

'Participation' has become an often meaningless buzzword (Blundell Jones et al., 2005, p.xiii) and is open to interpretation. The extent and implications of interaction and interchangeability between user/viewer and artist/architect is equally broad and ambiguous, having been criticised for an emphasis solely on process to the detriment of product and for a bias towards a collective good that omits authorship (Bishop, 2012; Miessen, 2011). Artistic or architectural participatory projects share the will to promote democratic ideals (Beebeejaun, 2016; pp.7–9; Helguera, 2011) and many participatory projects aim to empower their participants and inspire them to have a more active role in the making of their environments, becoming spatial agents and active placemakers rather than passive users, consumers or spectators (Schneider, 2013, p.250). These projects become political acts where, through direct action and activism, spatial agency is activated through collectivism (ibid.). Conversely, any romanticised nature of participation is contested as an 'un-questioned mode of inclusion that does not produce significant results, as criticality is challenged by the concept of the majority' (Miessen, 2011, pp.13–25). Participation thus often focuses on reaching a consensus, rather than on managing conflict, a processual weakness as reaching consensus does not represent a form of empowerment, but rather complacency when 'everything seems fine just as it is'. However, the state of 'being fine, often purely results from a lack of inquisitive thinking and the simple 'withdrawal from civic participation' (Sennett, 2013, pp.187–8). When supposed engagement does not occur, one questions this lack of interest, or apathy (Dürrschmidt, 2005). Apathy and lack of engagement are some of the difficulties encountered along the development of participatory processes that intend to incentivise collective, creative placemaking, challenging the assumption that participation is desired by all.

Outros Espaços intended to be a process where different and complementary views could be exchanged following Bakhtin's concept of dialogics (Holquist, 1981, p.291), where a discussion does not resolve itself in finding a common ground, but allows for complex situations to be analysed from different angles (Sennett, 2013, p.14–19) where conflict and consensus could coexist. Prior to a dialogue that would lead to the making of the project, the desire to make *Outros Espaços* materialise needed to be present amongst all participants:

> a schizoanalytical approach to 'participation' should start with desire, by considering the participative process as a way of assembling a collective economy of desire, articulating persons, gestures, economic and relational networks, etc. The participation process depends on participant's desire.
>
> (Petrescu, 2005, pp.43–63)

It was considered therefore that *Outros Espaços*, an initiative from Atelier Urban Nomads (AUN), was a desirable gift to all those who would participate in the project. Mauss (2002, p.3–9) argues that the exchange of objects between groups builds relationships between individuals. Such reciprocal gifting relates not only to a material gift, but also to political, economic, religious and social exchanges. However, *Outros Espaços* was not understood to be a gift under the same parameters as those described by Mauss. For AUN's gift to be of any value, it implied some form of reciprocity that should have materialised in clear communication, collaborative work and engagement and, importantly as Petrescu identifies, the desire of the community's participants.

Collaboration and cooperation are not interchangeable terms. The former implies action, whilst the latter relates to the act of enabling something to be done jointly. Cooperative encounters translate into an experience of mutual pleasure, a complex situation where people come together for a shared purpose, despite 'separate or conflicting interests', inequalities or lack of understanding of one another (Sennett, 2013, p.5–6). To cooperate requires skill, the ability to manage conflict and to shift from a passive to an active participatory presence (ibid., p.14). Collaboration or cooperation are not about establishing consensus but about making space for different views and opinions, for different perspectives to cohabit:

> We painfully discovered that collaboration is not about different disciplines and personalities climbing into a blender and producing a consensus. Rather, it has to be the deliberate creation of a sufficiently generous atmosphere to make room for the different disciplines and personalities, both ours and those of consultants, friends and lovers [. . .] being in one room, dialogues and eavesdropping inform projects.
>
> muf (2001, p.10)

These views of the role and value of participation for the making of urban environments leads to an unsettling position regarding whether or not participatory processes are worth pursuing as potential catalysts for spatial and social transformation and, consequently, as part of the process for the collective claim of one's right to the city (Lefebvre, 2003; Harvey, 2012). The 'freedom to make and remake ourselves and our cities' is 'one of the most precious yet most neglected of our human rights' (Harvey, 2012, p.4). The act of actively participating in the transformational process of our environments can reflect the engagement of citizens in claiming their right to the city as a collective entity, in relation to a political, economic, social and urban context. To dissociate the people from the governmental agencies that make decisions regarding the use and shape of the city would mean to create cities that are not designed for their inhabitants, but for speculative inhabitants fantasised by those in and with power, hence the importance of the citizens' active participation in the making of their urban contexts.

Even if not always idealised placebos for doing good and enacting democracy, writings on participation tend to emphasise the positive aspects of the practice, often omitting any difficulties encountered along the process. The acknowledgement that active participation leads to the making of more democratic environments assumes that a desire to participate is a given (Beebeejaun, 2016, p.8; Cooke and Kothari, 2011). Creative placemaking is considered to be a critical catalyst for the physical and social revitalisation of neighbourhoods, towns, cities and regions, animating public and private spaces (Markusen and Gadwa, 2010, pp.3–7). It promotes a decentralisation of the arts as a means to ignite a network of creative hubs and to democratise culture and both become associated with the development and regeneration of neglected or run-down urban areas. The danger lies in how these methods can easily become governmental tools that, through their social agenda, mask the roots for gentrification, 'urban cleansing', privatisation and the homogenisation of the public realm. They are no more than mere forms of entertainment and spectatorship rather than means to strengthen, create or empower communities for the making of their own built environments (Brenner et al., 2012, pp.24–41).

Project context and methodology

Outros Espaços took place in Beja II, a housing estate built in the late 1970s (after the 1974 revolution, Revolução do 25 de Abril) and is a ten-minute walk westwards from Beja's city centre. It was designated an ARU (Áreas de Reabilitação Urbana), categorised by having 'insufficient, run down or obsolete buildings and infrastructures; insufficient facilities for collective use; and lack of green spaces' (Governo de Portugal, 2013). Beja, the capital of Baixo Alentejo, a region in the south-east of Portugal, has 35,854 inhabitants and an ageing population (Governo de Portugal, 2013). It is a 'shrinking city' (Oswald, 2005), an area that faces depopulation as a result of economic decline and social problems reflecting symptoms of a structural crisis (Martinez-Fernandez et al., 2012). From 1981 to 2011, Beja's district suffered a demographic loss of 35,663 people, reflected in the number of derelict and abandoned spaces and building stock encountered across the district (Governo de Portugal, 2013). When built in the late 1970s the priority of Beja II was to provide housing facilities as there was a nation-wide urgency to re-house those who were living in very precarious conditions (Williams, 1981). The importance and value of public space was secondary and not acknowledged as a crucial element for the formation of a local community. Lack of engagement with the public realm was present since the neighbourhood's construction.

Outros Espaços was proposed by the author on behalf of AUN, a platform for the making of urban projects founded by the author in 2011. The proposal to Beja's Municipality (CMB) was in response to the ARU designation and intended to develop a collaboration with the local authorities, local non-governmental organisations, the local residents' association and a local school. The proposal aimed to overcome the problem of negligence and abandonment of shared public spaces in the neighbourhood. AUN's projects brought together transdisciplinary creative practices to create new collective spaces, grounded in a holistic design process built around site-specific, spatial and social narratives through visual storytelling. Each step of the design methodology was valued as much as the created spaces and each project presented a self-contained set of tools with a strong pedagogical agenda that aided the inhabitation and understanding of the intervention sites (Figure 11.1).

Contrary to common practice in Portuguese public realm projects, *Outros Espaços* did not aim to promote urban and environmental interventions to enhance life quality through a top-down approach. Rather, it intended to be inclusive and to work as a pedagogical tool to raise interest and awareness of Beja II's residents' regarding the importance of getting involved in the making and transformation of the non-existent public realm of their neigh-bourhood. *Outros Espaços* aimed to intervene in Beja II's Operações de

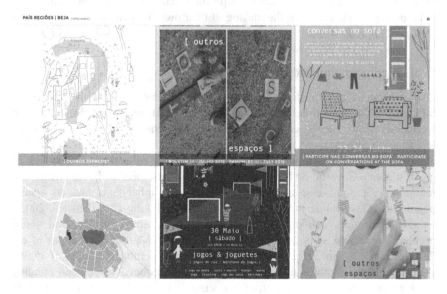

Figure 11.1   Outros Espaços project online archive.

Reabilitação Urbana (Urban Rehabilitation Operations) (ORUs) process with a more ethical and participatory approach in order to inform and complement the municipality's project procedures. This pedagogical approach was, however, a novel mode of working in Portugal and, consequently, unfamiliar amongst the project partners and, as the project evolved, it proved to be a challenge that needed overcoming for the engagement process to become more fluid.

ORUs resulted from the initial identification of certain urban areas as ARUs with the view of responding to identified problems. ORUs that involved not only the building stock but also the rehabilitation of infrastructures or green spaces were designated as Operações de Reabilitação Urbana Sistemáticas (Systematic ORUs) and benefited from a programme of public investment. These were covered by a Strategic Rehabilitation Programme which involved the submission of a design project to the IHRU (Instituto de Habitação e Reabilitação Urbana) for approval of the proposal by the municipal assembly and via public discussion with the local residents regarding the design and decision-making process. When approached by AUN, the municipality appeared to value and acknowledge the benefit of a more socially engaged and inclusive practice and *Outros Espaços* was perceived by the municipality as an opportunity to simultaneously respond to both spatial and social issues. However, for that to happen, collaboration was necessary, and for *that* to happen there was a need for an architecture practice that understood collaborative practice to be brought into the project: it was thought that without such understanding, the impact of this participatory and collaborative project may be compromised from its conception. It is important to note that the implementation of an ORU was not an initiative of Beja II's residents, nor was the project *Outros Espaços*, despite aiming to be inclusive of all parties. There was no 'collective economy of desire' (Petrescu, 2005, pp.43–63) to fuel the project and the transformation of Beja II's public spaces. Instead, *Outros Espaços'* role was to advise any decision that would affect the design of the ORU based on an informed involvement of the local residents and local organisations in order to prevent a bleak, non-site-specific design.

Similar to other participatory projects developed by AUN, *Outros Espaços* began with the setting up of the project team. This involved getting the local boroughs on board, along with a local school, local residents' association and local NGOs. Together these would become both the participants and the co-authors of the project. AUN's team would engage with the site and its inhabitants on a monthly or bi-monthly basis for periods of five to six days at a time with a site-based active mediator to maintain momentum between the visits. It was thought that clear and frequent communication would be crucial for the project to succeed as all management would be done from afar. There was no allocated budget for the development of *Outros Espaços*, the intention being to collectively and iteratively fundraise for the project's implementation. *Outros Espaços* did not have a defined outcome in place at the start of the project as the spatial, social and programmatic design would emerge from the discoveries found along the project's development process.

AUN worked with the students from a local school, on the principle that working with children and young people brings fresh, non-preconceived ideas to the design process and is a means of engaging younger people in thinking about their urban environments through creative and unconventional ways. It also forms part of a systemic approach; a way of reaching a wider group of people – relatives and friends – as the students would most likely share their makings with those they were close to. For *Outros Espaços*, AUN worked with a class of 14–16-year-old students of Escola Santiago Maior who were pursuing a professional path. The class was chosen based on their flexible timetable and on the number of students living in Beja II. AUN usually works with organisations based on-site that already have close contact with local residents facilitating the initial interaction in the period of time, during which the team remains in the background. This less prominent role of AUN at the initial stage leads to the identification of key residents who are more vocal within the neighbourhood and therefore, it is presumed, will help AUN approach those more reluctant to engage in something unknown and initiated by strangers, and time to develop these connections is key for the overall understanding of the value of AUN's projects. In *Outros Espaços*, the local authority and residents' association were the pivot for most other contacts within the neighbourhood.

At the core of *Outros Espaços* was the intention to reinvigorate Beja II's public realm by adding a social agenda that would give voice to the local residents before any proposal for the design of the public space adjacent to the residents' homes was to be developed further. By doing so, *Outros Espaços* specifically intended to prevent the construction of a homogenous non-site, nor user-specific, park. Passive users, consumers and spectators would be empowered in the project's process to take control over their environment by gaining some independence from the government, actively creating and building some of the public space around their homes. Unlike in projects where authorship is fully transferred from the artist or architect to the users, and the value of aesthetics supposedly becomes secondary, *Outros Espaços* promoted the transformation of the users' drawings and ideas into professionally resolved products and spaces. These would be part of a collaborative process that allowed different expertises to conjoin.

Based on AUN's methodological procedure, *Outros Espaços* involved six stages: preparing, observing, experiencing, experimenting, building and present-ing. All actions would happen outdoors, on-site, so that the project could start having a direct impact on the use of the space. The programme of action was outlined at the very beginning and shared with all involved. Preparing consisted of creating a logo for the project and an accessible website where the project's development would be documented as it evolved. Observing involved the inter-action with local residents and ethnographic research techniques, such as draw-ing, photography, film and mapping. Experiencing involved developing the observation process with the local residents, including interviewing. Experiment-ing involved the transformation of the collected imagery and ideas into new proposals and new design suggestions using methods such as stop-frame anima-tion and storytelling. Form as much as function was to be valued and neither form nor process should cause detriment to the outcome. Instead, all stages would have an equal weight, that being part of the holistic design approach of AUN. *Outros Espaços* would differ from relational art projects where process tended to surpass the value of form, and where a lack of ability to generate transformation in the long run was revealed to be problematic, as sharply criticised by Bishop (2012, p.51–79), leaving no legacy beyond the relational event(s). Building involved the construction of temporary (at an initial phase) small test infrastructures intended to respond to some of the identified problems experienced in the neighbourhood, such as lack of sport facilities and sheltered seating areas. Presenting was key for the success of the projects. To recurrently share the results of the process would be a means of demonstrating the importance of ones' involvement in the making of ones' own environment. It could take the form of a film projection, an exhibition or a meal, for example (Figures 11.2 and 11.3).

Quality and durability were also valued by AUN. These qualities were perceived as synonyms of pride and sense of ownership. Therefore, all designed elements of *Outros Espaços* intended to reflect quality and durability throughout the development of the participatory process. To produce something spatial and material was important as a first step for the eventual long-term transformation of the public spaces. As suggested by Potrč (SKOR Foundation, 2011), the value of making something together with people instead of merely talking would by

Figure 11.2 Escritório de Rua interviews with local residents.

Figure 11.3 Conversas no Sofá, interviews with and by Beja II's residents about the history and future of the neighbourhood.

itself start generating some form of spatial and political change. However, for a sense of pride to be felt, the quality of the objects or spaces being made was fundamental.

To better understand the participatory process of *Outros Espaços* and the obstacles encountered, two topics, collaboration and success, will be reflected upon.

Collaboration

In *Outros Espaços*, the desire to make room for an exchange of different thoughts, ideas and views to enter a dialogue and generate a participatory collaborative process never materialised. To participate implies some form of both collaboration and cooperation. However, the partnering organisations appeared to dismiss the possibility of a shared ownership of the project and, instead, relied on the complete control of the project by AUN. This was not part of AUN's role, which was proposed as that of initiators, not of off-site project managers. The agreement with the municipality was loose rather than written in the form of a contract, which led to differing interpretations of roles and responsibilities from the different parties involved. At the same time, neither was AUN's intention to adopt the mere role of facilitators, as criticised by Bishop (2006, pp.179–85) as being 'an attempt at the "elimination of author-ship", grounded in the anticapitalist premises and in a sort of Catholic altruism, a way to redeem the guilt of social privilege'. In an attempt to maintain author-ship, albeit collective, the project tried to involve multiple entities with different perspectives and backgrounds. The municipality and the residents would inevitably have very different views and intentions for the shared spaces of Beja II, as would the children and the local NGOs AUN worked with. However, for such a process to integrate these perspectives, some form of collaboration and interest would have to exist from all in the first place. Instead, there was a lack of commitment from the municipality towards the *Outros Espaços* that translated into apathy. This was interpreted as a lack of belief in the project by the residents, residents who had been previously deluded by CMB and who would not see AUN regularly enough to develop an understanding of the project's intent.

Triggering the desire to participate in the project was problematic. However, when *Outros Espaços* was initiated, the desire to collaborate seemed to be present, if not amongst all partnering organisations, then certainly by the municipality, which clearly manifested the will to embrace a joint venture for a shared good. However, it is possible that CMB did not fully acknowledge what collaboration implies and how crucial clear communication is for any form of collaboration to succeed. As for the residents, their previous awareness of how their opinions were usually dismissed or not even acknowledged by the munici-pality restrained the intention of *Outros Espaços* to become a platform for the residents to claim its ownership. Desire did not initially exist amongst the residents either. Lack of a collaborative attitude and poor communication were unable to confront lack of engagement and any existing apathy from all entities involved, including from AUN, who gradually started foreseeing the redundancy of its efforts to transform the neighbourhood's public realm. As the project evolved, AUN began to question the meaning of success in a participatory project. The initial expectations for *Outros Espaços* were replaced by the acknowledgement of the municipality's development of a non-inclusive design process for Beja II's public spaces.

Success

Success was important to AUN, firstly, to reassure the residents and the local associations that the project could have an impact, and secondly, for the municipality to justify the time and resources invested in *Outros Espaços*. An assumption that the project was unsuccessful derived from the limited involvement from the partnering organisations and lack of adherence from the residents in the different actions organised. These hindered the project's progression as it raised scepticism of its value. Collaboration, in this case, did not exist, nor were the neighbourhoods' public spaces socially or spatially transformed in a long-lasting way. It could thus be concluded that *Outros Espaços* should be considered unsuccessful. Yet, the meaning of success, in this context, is debatable.

Failure, according to Halberstam (2011, pp.1–25), can also become a means to question certain social standards that dictate the way certain people lead their lives. *Outros Espaços* opened a series of doors to question how public space is being designed in Portugal without the involvement of the users: how participatory projects, on their own, do not suffice, as the ground has not been prepared for them to actually happen or for them to have a long-term impact, leading to an overall lack of interest by all parties involved. Time is crucial for any participatory process to flourish. Without it, the necessary relationships and connections that will make the projects self-sustainable are less likely to have the opportunity to occur. *Outros Espaços* also allowed us to understand that the right to the city (Lefebvre, 2003) is not acknowledged either by the local authorities or by the residents. Despite supposedly being promoted by agents from the local authorities, the right to the city, as drafted in the *World Charter on the Right to the City* (Osorio, 2006, p.107), is not yet divested or practised even by those in the government. This reinforces the gap between the municipality and residents who were not being listened to and who, subsequently, do not expect their rights to be acknowledged, leading in turn to an overall attitude of apathy and indifference. Lastly, the involvement of children and young people in the process undermined the role of the architect to that of 'entertainer', as architectural pedagogy lacks strategies of participation and engagement in Portugal. The focus is instead on construction methods, urban planning, rehabilitation and maintenance and an architect's professional practice (Universidade de Coimbra, 2018; Universidade Lusófona do Porto, 2018; Escola de Arquitectura da Universidade do Minho, 2018).[1,2,3] The time, dedication, costs and social skills involved in this reform and in the participatory form of practice are perceived as unappealing and costly, yet it is increasingly promoted within urban agendas.

For all these supposed 'failures', *Outros Espaços* has shown that this work is needed so that a shift in attitude can potentially occur. For such to happen, the conventional understanding of success will have to be abandoned, or one would succumb to frustration. *Outros Espaços* can be considered successful, if the critique of its process is shared and used to pursue work that tackles a lack of involvement of residents and partnering organisations in the making of their living environment, fighting a lack of engagement experienced to date.

Conclusion: The unwanted gift – apathy and lack of engagement

The agreement with the municipality was loose rather than written in the form of a contract, which led to differing interpretations of the roles and responsibilities of the different parties involved. Neither the municipality, the other partnering organisations nor the residents appeared to have felt an urgent need for change to happen, or the lack of engagement would not have been experienced throughout the development *of Outros Espaços. Outros Espaços* was a 'gift' from AUN to Beja II's residents and to the municipality and other partnering organisations based on its altruistic motive to enhance the life and to empower those who lived the city. This gift was intended to trigger the residents' desire to be in control of their own environment so that their neighbourhood, and subsequently their city, could mirror who they are as individuals that share a physical space – Beja II's public space. It was a supposedly desired gift by the municipality, which, when presented with the projects' proposal, perceived it as an ideal complementary input to their intended strategy. However, it became no more than a placebo to address the previously identified social issues present in the neighbourhood.

Neither the project partners nor the residents had objectively asked for AUN's gift: *Outros Espaços* both revealed not to know how to make use of what was being offered and AUN made the assumption that the value of participatory projects for the making of public spaces was understood as a means to exercise the right to the city. Yet, its assumptions were wrong and AUN found itself assuming an altruistic role of gift giving. This phenomenon, however, is not new. The residents of Beja II, and other social housing estates, had been perceived as always being second in the municipality's list of priorities. There was a stigma attached to the neighbourhood for its history of drugs, vandalism and poverty, leading to its seclusion and containment within residential and familiar comfort zones, as established by the sociologist Neckel (in Dürrschmidt, 2005, p.275) 'emerg[ing] from the inner feeling that one has been personally degraded and exposed'. The gift AUN wanted to give was likely to have been received with a lack of interest and engagement and the 'altruistic' process of giving obscured such prediction. Before giving a gift, desire and an acknowledgement of its benefits ought to be triggered amongst the receivers.

Outros Espaços intended to promote a more democratic involvement of the residents in the making of their neighbourhood. In retrospect, it was trapped in its own romanticised idea of democratic placemaking, as criticised by Miessen (2011), despite its supposedly transformative agenda. *Outros Espaços* was never conceived as a placebo for democracy, but as a project with genuine intentions to 'co-create urban life for the better', so that the residents could demand 'their rights to be heard' and 'be involved more fully in the decision making' process (Beebeejaun, 2016, p.7). It was envisioned as an initiative that, although triggered by AUN, would eventually be taken over, appropriated and transformed by the residents and the projects' local partners, including the municipality. In turn, this transformation would benefit from an increase in proactivity by the residents contributing to a share of responsibilities between them, local NGOs and the municipality, which would be appealing to the latter. However, in practice, this would translate into the dismissal of the government's responsibility, as often occurs in government-funded art and architecture participatory projects which unwillingly become caught up in neoliberal agendas (Blundell Jones, Petrescu, Till, 2005).

Transformation was intrinsic in the design process, as was the idea of creating and designing through making, which intended to empower the residents and to contribute to generating autonomy from governmental agencies. If all had manifested as planned, *Outros Espaços* would have become an example of a participatory project that would contribute to raising awareness of ones' surroundings, both socially and spatially. *Outros Espaços* would not only reflect what already existed in the neighbourhood, but it would invite the materialisation of new ideas that would value and enrich the existing neglected spaces of Beja II. *Outros Espaços* is merely an example of a participatory project, which, despite its potential, had few, or possibly no, long-term results. With limited resources AUN tried to make the process creative and engaging, but that on its own proved not to be enough to generate the aims for social and spatial transformation that had been initially intended. To participate in the making of ones' urban environment and to be actively involved in the construction of the public realm is not solely the responsibility, or interest, of the individual or community that inhabits or uses public spaces. Instead, it ought to be a joint effort of the collective of individuals, organisations and the municipality as the representative governmental body. All of these need to be the participants, the makers of their cities.

Afterword

The phone rang. It was Renata's mother. Renata wanted to speak to me. She wanted to ask when we would be back, as she hadn't had so much fun in the neighbourhood as when we were around. Even if our aim was different, in the eyes of the residents our role was as 'animators', those who came to keep the children busy, entertained.

Drained by the process, apathy, lack of engagement and lack of collaboration we decided to temporarily withdraw from fieldwork. A moment of reflection was necessary to assess what had not succeeded as expected, to reflect on the incongruence of the initial journey. Some months later, when I got back in touch with the municipality, their 'yet to be drawn' project went to tender. I requested to see the package with the drawings. Unsurprisingly, I received no answer to the email I sent to the urban management department chief. After insisting, I was finally sent a series of Sketchup renders showing the future park. Amongst architects, as the name indicates, the software Sketchup was known for lacking precision and for being used to quickly represent 3D images that illustrate diagrammatic ideas. In this context, the Sketchup renders were being used as tender drawings and to promote the project. Had I not been told it was Beja II park, it could have been designed for any site, aimed at any user, negating all the work AUN did on-site and the municipality's supposedly socially minded agenda. Instead, the Sketchup proposal drawn by the municipality's architect was magically transformed into a built soulless park equal to any other place-less space. It involved no participation from anyone except for the municipality's urban management department team. Renata will not call the architect as she has never met him, and will never do so. *Outros Espaços* might not have succeeded in the way we had initially imagined. However, the lessons learned along the process are invaluable. Participation alone does not suffice to generate change, but now, more than ever, we believe that participatory processes developed in a creative way are necessary to trigger transformation at a local level. They can create places and connect people, though never by themselves, but as part of a bigger vision, a collective one (Figure 11.4).

Figure 11.4 Renata conducting interviews with the local residents.

Notes

1 Integrated Master's Degree in Architecture, 2018, Universidade de Coimbra, https://apps.uc.pt/courses/en/course/501.
2 Master Studies Architecture, 2018, Universidade Lusófona do Porto, www.masterstudies.com/universities/Portugal/ULP/.
3 School of Architecture, Universidade do Minho, www.uminho.pt/EN/uminho/Units/schools-and-institutes/Pages/School-of-Architecture.aspx.

References

Beebeejaun, Y. (eds.) (2016). *The Participatory City*. Berlin: Jovis Verlag GmbH.

Bishop, C. (2006). *The Social Turn: Collaboration and Its Discontents*. Artforum, February 2006 www.gc.cuny.edu/CUNY_GC/media/CUNY-Graduate-Center/PDF/Art%20History/Claire%20Bishop/Social-Turn.pdf [Accessed 23 April 2015].

Bishop, C. (2012). *Artificial Hells: Participatory Art and the Politics of Spectatorship*. London: Verso.

Blundell Jones, P., Petrescu, D. (2005). 'Losing control, keeping desire' in Blundell Jones, P., Petrescu, D., and Till, J. (eds.) *Architecture and Participation*. London: Spon Press.

Blundell Jones, P., Petrescu, D. and Till, J. (eds.) (2005). *Architecture and Participation*. London: Spon Press.

Brenner, N., Marcuse, P., and Mayer, M. (eds.) (2012). *Cities for People Not for Profit*. London: Routledge.

Cooke, B. and Kothari, U. (eds.) (2011). *Participation: The New Tyranny?* London: Zed Books.

Governo de Portugal. (2013). *Manual de Apoio : Processos de delimitação e de aprovação de Áreas de Reabilitação Urbana e de Operações de Reabilitação Urbana*. Ministério da Agricultura, do Mar, do Ambiente e do Ordenamento do Território. www.portaldahabitacao.pt/opencms/export/sites/portal/pt/portal/reabilitacao/ARUs/documentos/ManualdeApoioARU.pdf [Accessed 19 June 2014].

Dürrschmidt, J. (2005). 'Shrinkage mentality' in Oswald, P. (ed.). *Shrinking Cities Vol 1. International Research*. Ostfildern-Ruit: Hate Cantz Verlag.

Halberstam, J. (2011). *The Queer Art of Failure*. London: Duke University Press.

Harvey, D. (2012). *Rebel Cities: From the Right to the City to the Urban Revolution*. New York: Verso.

Helguera, P. (2011). *Education for Socially Engaged Art: A Materials and Techniques Handbook*. New York: Jorge Pinto Books.

Holquist, M. (ed.) (1981). *The Dialogic Imagination: Four Essays by M. M. Bakhtin*. Austin: University of Texas Press.

Lefebvre, H. (2003). *The Urban Revolution*. Minneapolis: University of Minnesota Press.

Markusen, A. and Gadwa, A. (2010). *Creative Placemaking: Executive Summary*. Washington: National Endowment for the Arts. www.arts.gov/sites/default/files/CreativePlacemaking-Paper.pdf [Accessed 3 June 2015].

Martinez-Fernandez, C., Audirac, I., Fol, S., Cunningham-Sabot, E. (2012). 'Shrinking cities: Urban challenges of globalization', *International Journal of Urban and Regional Research*, vol.36.2. http://onlinelibrary.wiley.com/doi/10.1111/j.1468-2427.2011.01092.x/pdf [Accessed 21 August 2016].

Mauss, M. (2002). *The Gift*. London: Routledge.

Miessen, M. (2011). *The Nightmare of Participation*. Berlin: Sternberg.

muf. (2001). *This is what we do, a muf manual*. London: ellipsis.

Osório, L. (2006). 'World charter on the right to the city' in Sané, P. and Tibaijuka, A. (ed). *Débats publics internationaux: Politiques Urbaines et Le Droit à la Ville*, Unesco. http://unesdoc.unesco.org/images/0014/001461/146179m.pdf. [Accessed 3 June 2015].

Oswald, P. (2005). *Shrinking Cities, Vol 1. International Research*. Ostfildern-Ruit: Hate Cantz Verlag.

Petrescu, D. (2005). 'Losing control, keeping desire' in Petrescu, D. and Till, J. (eds.) (2005). *Architecture and Participation*. London: Spon Press.

Schneider, T. (2013). 'The paradox of social architectures' in Cupers, K. (ed.). *Use Matters: An Alternative History of Architecture*. New York: Routledge.

Sennett, R. (2008). *The Craftsman*. London: Penguin Books.

Sennett, R. (2013). *Together*. London: Penguin Books.

SKOR Foundation. (2011). *Interview with Marjetica Potrč*. Available at: www.youtube.com/watch?v=B2M0qxHcYfc [Accessed 18 May 2017].

Universidade Lusíada do Porto (2018). Available at: www.por.ulusiada.pt/cursos/1ciclo/1ciclo.php?cp=012. [Accessed 21 January 2018].

Universidade do Minho (2018). Available at: www.arquitectura.uminho.pt/pt/Ensino/Licen ciaturas_e_Mestrados_Integrados/Mestrado-Integrado-em-Arquitetura/Paginas/Plano-de-Estudos.aspx [Accessed 21 January 2018].

Williams, A. (1981). 'Portugal's illegal housing', *Planning Outlook*, 23.

Section 5

Extending Ecologies

12 Towards beauty and a civics of place

Notes from the Thriving Cities Project

Anna Marazuela Kim and Joshua J. Yates

Abstract

This chapter proposes a new paradigm to address some of the central challenges of the creative placemaking movement: a framework of urban thriving based upon a model of human ecology. We introduce the paradigm as it emerged over five-years of research and practice at the Thriving Cities Project; then highlight the role of 'beauty' within this framework, in its potential to engender attitudes of care and commitment foundational to civic agency. In conclusion, we suggest that the intersection of aesthetics and civic agency – conceptualized in terms of placemaking as a 'civics of place' – as a promising area for future research.

Introduction

Now that the creative placemaking movement has received broad-based support for nearly a decade, it can begin to reflect upon one of its more ambitious aims: to enact resilient and effective processes of citizen-led placemaking (Markusen, 2014). We have ample evidence of the remarkable, but at times temporal, power of arts-based interventions in communities. The challenge is whether we might effect structural processes and conditions to create, and also sustain, more enlivened, equitable places of inhabitation. Given the current state of social disparity and political life in the US and the UK, where creative placemaking has had the longest history, critical attention to this agenda seems not only pressing, but also opportune. Among the questions we might pose is how to cultivate civic engagement in urban communities when its inhabitants – whether mobile, displaced, or subject to social exclusion – may feel little or no connection to place (Bedoya, 2013). In addition, what arguments can we bring against the displacement that at times follows from gentrification, when success in urban regeneration is still judged primarily in economic terms? This chapter proposes a new paradigm to address some of the central challenges of the creative placemaking movement: a framework of urban thriving based upon a model of human ecology. In the first part, we introduce the paradigm as it has emerged over five years of combined research and practice at the Thriving Cities Project.[1] In the second, we highlight the role of 'Beauty' within this framework, in its potential to engender attitudes of care and commitment foundational to civic agency. In conclusion, we suggest that the intersection of aesthetics and civic agency – conceptualized in terms of placemaking as a 'civics of place' – as a promising area for future research.

Why 'thriving'?

We begin with the acknowledgement that neither of us arrive to these questions from the usual backgrounds or training that characterize the field, but rather from a shared interest in larger questions of meaning pursued within the interdisciplinary community of the Institute of Advanced Studies in Culture[2] at the University of Virginia, where the Thriving Cities Project was initially launched. As such, the project has been shaped from its outset by an aim central to this particular intellectual community: to articulate richer conceptions of human thriving, as it is challenged under the conditions of globalized or advanced capitalism.

At the heart of the Thriving Cities Project is the question of what constitutes human thriving, considered within the specific context of cities and civic activity. In taking 'thriving' as its conceptual starting point, our project reflects the re-orientation of ethics, from utilitarian or quantitative measures of the 'good', towards a more capacious, Aristotelian idea of flourishing (*eudaimonia*), which was also envisioned within the context of the *polis*. While not adopting an Aristotelian position, we take 'thriving' roughly to mean the realization of human capabilities, as philosophers such as Nussbaum (1988, 2000, 2011) have theorized, and also acknowledge the historic and present circumstances particular to each community in which those capabilities are constrained or realized. In building upon a philosophical tradition that has importantly shaped not only ethics but economic theory and policies of world development (Sen, 1999), we hope to contribute more than a 'fuzzy concept' to the placemaking debate (Markusen, 2012; Gadwa Nicodemus, 2013). For the Thriving Cities Project, the concept of thriving also works on a practical level, to shift thinking from conventional deficit perspectives to one of asset orientation, encouraging leaders and citizens to see beyond common problems to collective possibilities. This double intention – to combine theory and practice, testing one against another – serves as a guiding principle throughout. Briefly, over the course of its initial development, the aim of the Thriving Cities Project has been threefold: to formulate a more complex, holistic picture of what thriving is, both conceptually and for a particular city or urban environment; to apply research towards the question of how this thriving might be fostered, and test this research against the practical experience and knowledge of a diverse and unconventional range of stakeholders, in collaboration with eight pilot cities in the US; finally, to develop effective, user-friendly tools for the practitioner, from grassroots to government, to advance the thriving of their individual community.

An ecological model of cities

Our focus on the urban dimension of thriving is motivated by the exponential rise of cities and their significance for the future of the planet. It is estimated that by 2050, the urban population of the world will nearly double (UN-Habitat, 2017, p.3). In this shift, cities will become the grounds for many of the consequential challenges humanity faces – demographic, ecological, technological, economic, political, and social. Cities will likewise become the principal forging grounds for solutions to these challenges in the years ahead. What it means and takes to thrive in urban environments in the twenty-first century – what will count as societal and material progress – is arguably one of the great challenges of the present era, a challenge that will be as much cultural and ethical as it is economic or technological. We live in a time of opportunity for advocating for the role of the arts in this development, but more importantly to define and press for the conditions, structures, and kinds of commitments that will bring about greater equity and thriving for greater parts of the population. Towards that end, the first aim of the Thriving Cities Project has been to develop a new paradigm for understanding and assessing what constitutes success in these terms.

At the foundation of the paradigm is the recognition that cities are complex, asymmetric and dynamic social environments that are constantly changing and evolving. Here we build upon Jacob's idea of cities as forms of 'organized complexity,' as developed in her increasingly acknowledged study, *The Death and Life of Great American Cities* (Jacobs, 1961). Much like biologists think of an ecosystem as a community of living and non-living things working together in the natural world, Thriving Cities adopts a human ecology framework to help envision the interrelated structure and workings of a city. An ecological model brings into view the idea that cities are neither collections of autonomous individuals or discrete problem areas, like poverty or affordable housing, each hermetically sealed from one another; nor do cities behave like mechanical systems that can be managed and controlled by rational experts in a top-down manner. In this regard, it also shares the principles of models of new urban governance, which stress democratic, participatory networks of stakeholders across all levels (UN-Habitat, 2016; Brain, 2005). The ecological approach helps reveal the unique features of a city as a social organism that both empowers and constrains the ways of life and life chances of its residents. Finally, it insists that the human ecology of a city is specific to the social, historical, and moral context of a particular place that gives each city its own unique and ever-evolving identity.

The Endowment Framework: An ecological paradigm for civic thriving

In order to make sense of the human ecology of a given city, and provide tools to empower communities to use their distinctive history, resources, and culture to foster greater thriving, Thriving Cities has developed an Endowment Framework[3] organized around six realms of civic life. The language of Endowments is highly intentional. It stands in dynamic tension with the language of capital used by most standard and many cutting-edge approaches. Where 'capital' denotes abstract, a-temporal, and amoral value that is at once fungible and fluid (which is to say unfixed – precisely the source of its conceptual strength), the language of Endowments brings the dimensions of particularity and temporality back into view. Endowments are the products of investments made over time; they must be maintained in the present if they are to remain available in the future. Also implicit in the language of Endowments is a sense of fiduciary responsibility and obligation. Where capital functions as a medium of value and exchange irrespective of context, Endowments function as a reservoir of wealth held in common, as a trust within very definite contexts. Despite its obvious strengths, the language of capital is not able to capture these essential qualities of community life. Not surprisingly, it remains empirically elusive in approaches that rely on it.

We define 'Endowment' as a realm of human activity that generates socially determined value and ends. The generative value produced within an Endowment is realized by and through existing resources comprised of cultural practices and vocations, perceptions and ideas, and institutions. This is a distinctively cultural approach, one that emphasizes the normative dimensions of common life in cities and invites us to see them in terms of six interactive (and ever-evolving) formative contexts in which we routinely see the exercise of moral agency and practical reasoning across human communities. To gain analytic purchase on the interconnections that comprise the human ecology of a city, and to measure them for assessment, we articulate that interconnecting web in terms of six dimensions. We call these 'community Endowments.' As with ecologies in the natural world, human ecologies contain and depend upon an array of fundamental Endowments (i.e., capacities and qualities as much as quantities). In the case of human ecologies, such Endowments: give expression to long-standing and universally recognizable ends that are essential to human thriving (e.g., intellectual life, aesthetics, sociality, play, health, security, and transcendence); become actualized within specific social practices and institutions (e.g., universities, theaters, social media, clubs and associations, hospitals and clinics, and places of worship); have distinctive histories that shape their present and future possibilities; and interact dynamically with one another, creating both virtuous cycles when robust and healthy, and vicious cycles when depleted and weak, but also generating synergies with unintended consequences and tensions between competing goods.

Taking again a philosophical view as our conceptual starting point, the first three of the six Endowments build on the classical ideals of 'The True,' 'The Good,' and 'The Beautiful,' which serve as placeholders for our initial attempt to capture the larger dimensions of human thriving. The last three are what we might call modern ideals of 'The Prosperous,' 'The Just,' and 'The Sustainable.' Together they describe some of the most recognizable horizons of human experience, and the building blocks of thriving in any given community.

'The True'

The Endowment of 'The True' contains realms of activities that are expressed through educational institutions and in informal settings. These realms generate the value of knowledge for a given city, many of which are investing in knowledge and innovation districts. Resources and indicators for this area might include the quality of education available there; access to remedial training; the number of media outlets; vocational training centers; and numbers of educated citizens, teachers, and tutors. The contextual realities that constrain and enable this Endowment within a city can include the politics of education funding and burgeoning social media platforms.

'*The Good*'

The Endowment of 'The Good' is comprised by activities that are expressed through philanthropy, religion, and non-profits. Resources that constitute this Endowment might include civic-minded organizations, volunteer hours, and geographical proximity and overlap of different populations. Equally important for this Endowment are areas in a city that allow for the creation of different relationships, such as city sports leagues and public spaces. The contextual realities that constrain and enable this Endowment within a city can include demographic trends, decline of traditional institutions, forms of individualism, legacies of racial distrust, and increasing diversity.

'*The Prosperous*'

The Endowment of 'The Prosperous' encompasses the activities that are expressed through businesses and personal and social thrift. These realms gen-erate the value of economic wealth for a city. Increasingly, cities are being viewed as central places for economic growth due to certain structural dynamics associated with urban places, such as industry clustering, agglomeration or density of networks, and attraction of skilled labor. With the growth of the knowledge economy, these particular urban features will become more salient as economic growth will largely consist of technological innovation. Other aspects of this Endowment assess the fiscal health of the city, the tax burden on residents and businesses, and the affordability and availability of housing stock. Equally important are the structures of economic opportunity within cities that highlight the distribution and quality of jobs, access to job training, and ease of credit.

'*The Just*'

The Endowment of 'The Just' is characterized by activities expressed through government, advocacy groups, and justice institutions. These realms generate the value of public justice and political/civic order. The primary realm of this Endowment is government institutions that ensure the smooth and just manage-ment of the city. The resources in the US might include the Board of Alderman, the city treasury, and various municipal courts. In addition to public administra-tions, private actors such as advocacy groups, lawyers, and political activists play important roles in this Endowment. Resources that constitute this aspect of the

Endowment are perceptions of community safety, levels of corruption, and voting levels. By focusing on issues of justice, this Endowment also highlights the larger inequitable power relations and unjust spatial distribution of resources based on race, culture, or class. The contextual realities that constrain and enable this Endowment within a city can include sectarian politics, legacies of racial injustice, and limits imposed from federal and state agencies.

'The Sustainable'

The Endowment of 'The Sustainable' contains realms of activities expressed through resource management, biotic care, and public health. These realms generate the value of environmental and biological health of a city and its inhabitants. A key issue for a city is how well it stewards important natural resources. This includes both its efficiency and resiliency. Resources associated with this realm of activity include public works (excluding water management and recycling centers) and private organizations, such as environmental advocacy groups. For biotic engagement, resources that facilitate opportunities for residents to interact with and care for the natural world, such as riverfronts or outdoor ventures, are included. Finally, public health captures proximity to healthcare, access to fitness centers, and prevalence of diseases within a community. The contextual realities that constrain and enable this Endowment within a city can include legacies of urban pollution, physical geography, and environmental regulations.

The case for 'The Beautiful' as an Endowment of thriving cities

Beauty would seem to be a key indicator of a city's thriving. Yet not only is it difficult to define and measure, it also stands in need of some defense. Beyond the aesthetic delight we take in a well-designed city and its amenities, beauty's larger significance may be indicated by pointing to what results in its absence. As evident in slums and other forms of urban blight, human flourishing seems intimately bound up with the built environment and the possibilities it affords. Yet when 'beauty' is ranged against other Endowments, especially those that seem to more directly address fundamental aspects of thriving – education, justice, and economic wellbeing – its significance recedes from view. Much like the arts, which are among the first programs to be cut when budgets are tight, beauty can seem a mere luxury at the top of a structure whose foundations desperately need attention, or worse: a value that obscures or distorts our very attention to those needs.

Moreover, historically, beauty, like the values of truth and goodness with which it is traditionally allied, has been subject to competing interpretations and claims that at times seem incapable of resolution. This would seem problematic if we, or various constituencies of a given city, are to invoke it as a measure of what it means to thrive, and moreover to claim it as a source of unified vision, rather than division. Any attempt to recover a productive concept of beauty must proceed from the awareness that it is philosophically and culturally complex in these and other ways. In our initial research Brief (Kim et al., 2014)[4] to make a case for 'beauty' in an expanded frame, as architectural theorists such as Meyer have proposed (2015), we directly acknowledge the challenges posed by our Endowment, but also its potential. For the purposes of our Brief, we define the realm of 'The Beautiful' to encompass the built environment and urban design crucial to the infrastructure of the city and our experience of it. We also highlight the role of the arts, at the level of community, in the health and vitality of the urban commons, particularly in their potential to foster imaginative responses to urban challenges and dialogue across societal boundaries. More broadly, we consider 'beauty' in terms of the aesthetic orientation fundamental to human life, and its capacity to foster attitudes of care for the urban commons. Within the discussion of these domains, we draw out the contribution of 'The Beautiful' to thriving in its dynamic relation to other Endowments, in particular 'The Good,' 'The Just,' 'The Prosperous,' and 'The True'. Casting the realm of 'The Beautiful' in this larger ecology of forces not only highlights its catalyzing role, but, more importantly, offers critical leverage against any one-sided, positive valuation of processes with which it is at times allied, such as beautification and gentrification.

The arts have often been valued as sites for cultivating our orientation towards beauty and shared experiences that foster communal imagination and vision. But it is an indication of the resourcefulness of the arts that they also function as a source of autonomous vision and critique. In their re-presentation of reality, the arts, in their manifold expression, are also pockets of reflective resistance against society's mainstream values of consumer culture, profitability, and even beautification. Here we embrace the paradox that the realm of 'The Beautiful' encompasses that which seems to subvert traditional notions of beauty, affirming that 'The Beautiful' has the potential to provide alternative, transforming images to those of oppression. From this perspective, we might think of beauty as it cultivates a fundamental orientation, attunement, or awareness that enables us to experience reality in new, dynamic, and potentially disruptive or even disturbing ways (Meyer, 2015, pp.32–4). Beauty encourages us to bracket, if even for a moment, our everyday mode of being in the world, transforming our experience of it – and potentially ourselves. Beauty arguably has an important social dimension as well. Not only is it an experience we share or generally want to share with others, as the history of art, literature, and other forms of human expression attest, but experiences of beauty might also be described as particularly intense forms of communication, connecting people in communities of appreciation. As the philosopher Nehamas (2007, p.82) writes: 'What is involved is less a matter of *understanding* and more a matter of hope, of *establishing* a community that centers around it – a community, to be sure, whose boundaries are constantly shifting and whose edges are never stable.' At the societal level, whether these communities are uniform or diverse will depend upon the context in and conditions under which beauty is enacted and mobilized. While there is ample evidence of beauty's power to create exclusive communities of 'taste,' beauty also has the potential to create unusual coalitions among otherwise polarized, segregated, or inequitable populations, in the pursuit of higher or intrinsic values.

Beauty's social architecture: Art as a public good

From this point of departure, we begin to grasp how 'beauty' is a phenomenon that, even at the level of its aesthetic genesis, holds important possibilities for individual experience and the social order. But what about the place of these possibilities within the practical forces of urban life – what we might call the architecture of the body politic? What has beauty to do with the tangled yet vital nexus of social macrostructures, civil society, and cultural policy? How do the arts' sectors of our cities stand within the push and pull of proprietary markets, government agencies, and nonprofit organizations? This is a large question that concerns scholars and practitioners alike, and cannot here be addressed in full. However, some compelling indicators for advancing our thought can be found in studies such as Zuidervaart's (2010) *Art in Public: Politics, Economics, and a Democratic Culture*. Following Zuidervaart, we stress two points of orientation that help contextualize the Endowment of 'The Beautiful' along these lines: first, the nature of art as a public good in civil society; and second, art as an autonomous yet interconnected form of healthy participation in the social economy. These emphases also reveal how 'The Beautiful' intersects more tangibly with the Endowments of 'The Just' and 'The Prosperous.'

Zuidervaart's systemic study contributes a useful framework for conceptualizing art as a public good. Zuidervaart begins by articulating the limitations that contemporary market-based economic analyses and instrumentalist, political models of liberalism bring to bear on this issue. Here our understanding of a social 'good' must first overcome a 'cultural deficit' prone to evaluating art on the basis of benefit principles alone, and a 'democratic deficit' prone to understanding administrative state involvement in the arts on the basis of an insufficient conception of public justice. These concerns in many ways echo and specify a tension between a rational-scientific point of view and the more broadly aesthetic orientation described above. Both deficits highlight the need to reposition the question of aesthetic value within a larger reflection on institutional and cultural pluralism and the goals of a deliberative democracy. One way forward is captured in Zuidervaart's attention to the macrostructure of 'civil society.' The term denotes a site of vital interface between the proprietary economy and the administrative state, as well as that constitutive element of democratic communication crucial to mediating the meaning of public goods. It is here that we find one normative dimension for understanding how the arts are societally important (ibid., p.91), and thus a specific resource for situating beauty more conspicuously in a trajectory of public thriving.

Civic participation and the arts

Thinking in terms of the public character of art has to do with identifying how the body politic needs the arts and the arts need the political and economic systems that, all things considered, too often undermine creative communication and exert systemic pressures on arts-related initiatives. Prioritizing the arts in society requires the lead players (institutions, initiatives, and decision-making bodies) to recognize one another's contributions and to exercise together a committed participation in the life of the public realm and goods. The point combines the better sides of democratic idealism and pragmatic public policy. The case is not just for the good of art, but for a healthy transformation of culture. Towards such ends, we need to consider how the arts indeed strengthen civil society and merit more genuine participation in the social economy. Doing so in turn requires us to consider 'participation' with respect for the work of 'nonprofit, mutual benefit, and nongovernmental organizations' (Zuidervaart, 2010, p.132), and how their work on behalf of the arts happens in relation to the proprietary market and government organizations. Zuidervaart acknowledges how the mission-driven character of civic sector organizations might not always comport with the capitalist criteria of the proprietary market, but says that this need not weaken the call to think of art and its public in terms of societal and structural transformation (ibid., p.151). Mutual respect and recognition can go a long way. The proprietary economy needs to mitigate its more reductive models of participation, and arts organizations need to promote 'artistic practices and relations that are sociocultural goods' (ibid., p.168).

Alongside these points, it is important to remain sensitive to the balance between artistic autonomy and social inter-subjectivity. The balance parallels an important dynamic in our practical philosophical points about how artistic beauty works. In terms of aesthetic production, what the artist puts into a work is realized most fully in the event of aesthetic experience, in which viewers engage with the work. In terms of a social architecture, the arts indeed enjoy an internal autonomy, but their underlying dimensions are always already interpersonal: as processes, products, and events, they involve multiple publics. The authenticity of any resulting participation, for Zuidervaart, depends upon a delicate balance between intrinsic artistic worth and extrinsic practices of imaginative disclosure. In other words, what individual artists and arts sectors both do inherently involves an outward movement. Here, drawing on the work of Taylor, Zuidervaart speaks about the compatibility of artistic authenticity and social responsibility within a democracy. The point parallels the ethico-political trajectory of aesthetic experience highlighted by philosophers such as Gadamer (1986). If art is a social institution, then the goal of 'imaginative disclosure' (in, for example, public art in new genres) necessarily involves the responsible contribution of collaborators, critics, and publics. The 'co-responsibility in the creative process' (Zuidervaart, 2010, p.263) in this sense concerns the place of the arts' sector within an urban society, as well as the more particular place of a given work of art within an individual's experience of beauty in that society. All told, Zuidervaart equips us to discern how beauty, as manifested in artistic practices and priorities, 'can foster the growth of democratic culture' (ibid., p.269). Understanding the public character of art is not about proving the bottom-line profitability of museums and galleries, performances and festivals, nor is it about extending special entitlements to appease constituencies. It is about appreciating how the arts are poised in a unique way to serve the common good.

Contextualizing 'The Beautiful' as an asset to urban thriving in many ways centers on the idea that the artistic and creative vitality of cities can yield a pronounced impact on the social – and not just economic – strength of urban life. It also concerns us – as individuals, precisely in the context of our lived sense of who we are, how we live, and with whom we are intrinsically connected. In our Brief, we explore, by way of indicators drawn from practical philosophical considerations, a macro-societal conception of art as a public good and research paradigms that advance attention to the cultural and community value of urban arts sectors. Each of these touchstones speaks uniquely to the aesthetic architecture of a city, and to the manner in which beauty is a practice through which we find and make our place amid the physical and social landscapes of urban life. We focus in this Brief on the role of the arts in this itinerary towards Creative Placemaking, in necessarily technical terms. But aesthetic beauty, we also suggest, does not always accommodate itself to formulaic calculations or benchmarks, and sometimes shows recalcitrance before reductive definition. Yet such is the wonder, the appeal, and the urgency of the arts. Though the aesthetic architecture of a city is not an easy matter to encircle and appraise, seeking out the places and practices of 'The Beautiful' across our neighborhoods and broader networks allows us to engage and affirm our urban imagination all the more in the service of the higher aims of the Creative Placemaking movement.

Future research: From Creative Placemaking to a civics of place

While the arts are championed as a neutral resource with the potential to bring together and transform diverse constituencies, arts-based initiatives are not immune to perceptions of bias, privilege, and insensitivity towards pressing issues of social injustice and exclusion. As one line of response, we propose that a central, ethical challenge of creative placemaking might be addressed through the paradigm outlined above: an ecologically-based framework that articulates and emphasizes the necessary interrelation and contribution of a city's cultural Endowments, particularly the realm of 'The Beautiful,' 'as it plays' a central and catalyzing role, but is in turn responsive to and constrained by other domains and values.

While the question of what constitutes beauty in a particular culture or society is open to debate, we believe that it is both possible and necessary to become more articulate in regard to beauty. In its potential to mobilize resources of an imaginative, and at times critical and transformative, vision of what a city and its citizens might be, we believe beauty also has the potential to bridge disparate social, economic, racial, and political divides in a time when initiatives of this kind are critically needed. With regard to beauty and the city, we find it significant that major, recent public surveys invoke beauty as precisely that value which might serve as a common ground of consensus building among diverse populations: the UK Government's *People and Places: Public Attitudes to Beauty* (2010) and in the US, studies such as *Beautiful Places: The Role of Perceived Aesthetic Beauty in Community Satisfaction* (Florida et al., 2009). Beyond the renewed emphasis in many studies on good urban design and greater access to aesthetic spaces, there are signs that beauty is returning to areas of discourse on urbanism from which it had been largely displaced, for example, by an overriding concern with issues such as sustainability. As the landscape architect Meyer (2008, p.6) argues in her 'manifesto,' sustainability itself might be seen to depend upon the re-orientation of consciousness that beauty and aesthetic experience fosters: the cultivation of commitment and care that depends upon a shift from 'an egocentric to a more bio-centric perspective.' In terms of beauty's 're-orienting' capacity, one might also pursue parallels to what is described by theories of biophilic response, which, on one level, may be interpreted as a displacement of the aesthetic into the realm of ecological affiliation.

Along similar lines, across many urban communities today, the aesthetic dimension of beauty intersects with initiatives that deliberately involve artistic practices in order to cultivate a richer consciousness of the public good. This intersection of aesthetics and civic agency, conceptualized in terms of placemaking as 'civics of place,' offers a promising area for future research. Placemaking, in its varied dimensions and possibilities, captures a significant array of functions along the continuum of human ecology. It speaks to the most fundamental need of individuals to have a place in the world, as part of a community that supports their full placemaking potential, and speaks as well as to the creative activity by which this is made possible. The art of placemaking reveals the dynamic way in which beauty can operate as a coefficient that multiplies the creative and caring endeavors of citizens and institutions committed to promoting the public good. To envision this possibility, we will need to better express and capture how a city's aesthetic orientation can support a fuller consciousness of an individual's place in a given community, and also a richer mode of engaging the common good. While daunting, to this end we might draw upon a recent gathering on Beauty and Justice to which Thriving Cities was invited, as part of the Ford Foundation's yearlong initiative, *The Art of Change*:

> Beauty is not essential because it is in service to some economic or social outcome that is more basic than it; it is in and of itself a basic need. Together we will investigate how we might more effectively articulate, value, and nurture beauty as a basic need and right in our society ... We will consider the interdependence between beauty and justice, exploring how beauty itself is a kind of justice, and also how beauty can be an agent of justice because of what it provokes in us. We will ask: How might we leverage this inherent yet often unexamined connection between beauty and justice in order to build, in the words of Ford Foundation president Darren Walker, 'an economy of empathy'?
>
> (Frasz, 2015, n.p.)

Beyond articulating the significance of beauty, how might we move from creative placemaking to the construction of actionable, sustainable civic processes and infrastructure? This practical question has been a driving concern of the Thriving

Cities Project over the course of its initial development. It began by creating a new paradigm to identify and better describe the issues at stake; then to proof of the validity and usefulness of this framework through conversations with stakeholders and practical application in the field, according to which it was also rethought. The first idea of what was crucial to thriving we called 'civic substructure,' a notion that has origins in the idea of place as root and foundation. This concept was next expanded to think in terms of two interrelated ideas: civic capacity and civic infrastructure. Capacity building in particular figures centrally in calls for a new urban agenda, in HABITAT III's 2016 Policy Paper, and in Americans for the Arts' report on arts and cultural districts (Ashley, 2014).

After the completion of its final year as an experiment between research and practice, the Thriving Cities Project moves into its next phase as a collaborative partnership between two new entities. The Thriving Cities Lab,[5] a research outpost at the Institute for Advanced Studies in Culture, aims to understand the nature of our epochal transformation into urban inhabitants; critically engage its consequences for human thriving; and equip scholars and communities with the intellectual resources necessary for constructively meeting the demands of this grand challenge. The Thriving Cities Group,[6] a not-for-profit Civic Ventures Firm, will work to translate the Lab's research into tools and services to practically equip residents, practitioners, investors, and experts over the next decade to rebuild the civic infrastructure of their cities and regions so all can thrive. Together, these two entities will continue to engage with questions about the nature of our moment and the possibilities of urban thriving, working at the intersection of research and practice.

Notes

1 Thriving Cities Project, http://iasculture.org/research/culture-capitalism-global-change/thriving-cities
2 Institute for Advanced Studies in Culture, http://iasculture.org/
3 Thriving Cities Endowment Framework, http://thrivingcities.com/endowments
4 Endowment Brief on the Beautiful, http://thrivingcities.com/endowments/beautiful
5 Thriving Cities Lab, http://iasculture.org/research/thriving-cities-lab
6 Thriving Cities Group, www.thrivingcitiesgroup.com/

References

Ashley, A. J. (2014). *Creating Capacity: Strategic Approaches to Managing Arts, Culture, and Entertainment Districts' in Americans for the Arts*. Washington, DC: National Cultural Districts Exchange. [Online]. Available from: www.americansforthearts.org/sites/default/files/pdf/2014/by_program/reports_and_data/toolkits/cultural_districts/issue_briefs/Creating-Capacity-Strategic-Approaches-to-Managing-Arts-Culture-And-Entertainment-Districts.pdf. [Accessed: 2 January 2018].
Bedoya, R. (2013). 'Placemaking and the Politics of Belonging and Dis-Belonging', *Grantmakers and the Arts Reader*, 24: 20–21, 32. [Online]. Available from: www.giarts.org/article/placemaking-and-politics-belonging-and-dis-belonging. [Accessed: 2 January 2018].

Brain, D. (2005). 'From Good Neighborhoods to Sustainable Cities: Social Science and the Social Agenda of the New Urbanism', *International Regional Science Review*, 28(2).

Florida, R., Mellander, C., and Stolarick, K. (2009). 'Beautiful Places: The Role of Perceived Aesthetic Beauty in Community Satisfaction' in *Working Paper Series: Martin Prosperity Research* [Online]. Available from: www.creativeclass.com/rfcgdb/articles/Beautiful%20places.pdf. [Accessed: 2 January 2018].

Frasz, A. (2015). 'Is Beauty a Basic Need and Right?' in *Helicon*. [Online]. Available from: http://heliconcollab.net/beauty/. [Accessed: 2 January 2018].

Gadamer, H. G. (1986). *The Relevance of the Beautiful and Other Essays*. Robert Bernasconi (ed.), Nicholas Walker (trans.). Cambridge, England: Cambridge University Press.

Gadwa Nicodemus, A. (2013). 'Fuzzy Vibrancy: Creative Placemaking as Ascendant US Cultural Policy', *Cultural Trends*, 3-4.

Harvey, A. and Julian, C. (2015). 'A Community Right to Beauty: Giving Communities the Power to Shape, Enhance and Create Beauiful Places, Buildings and Spaces' in ResPublica. [Online]. Available from: http://www.respublica.org.uk/wp-content/uploads/2015/07/Right-to-Beauty-Final-1.pdf. [Accessed: 2 January 2018].

Ipsos, M. O. R. I. (2010). 'People and Places: Public Attitudes to Beauty.' [Online]. Available from: www.ipsos-mori.com/researchpublications/publications/1668/People-and-places-Public-attitudes-to-beauty.aspx. [Accessed: 2 January 2018].

Jacobs, J. (1961). *The Death and Life of Great American Cities*. New York: Vintage.

Kim, A. M., Yates, C.S., Merrill, E., and Miller, C. (2014). 'Thriving Cities Endowment Brief: The Beautiful' Institute for Advanced Studies in Culture, Charlottesville, Virginia. [Online]. Available from: http://iasculture.org/research/publications/beautiful-endowment. [Accessed: 2 January 2018].

Markusen, A. (2014). 'Creative Cities: A Ten-Year Research Agenda', *Journal of Urban Affairs*, 36(S2).

Markusen, A. (2012). 'Fuzzy Concepts, Proxy Data: Why Indicators Won't Track Creative Placemaking Success' in *Createquity*. [Online]. Available from: http://createquity.com/2012/11/fuzzy-concepts-proxy-data-why-indicators-wont-track-creative-placemaking-success/. [Accessed: 2 January 2018].

Meyer, E. K. (2015). 'Beyond 'Sustaining Beauty' Musings on a Manifesto', in Deming, M. E. (ed.), *Values in Landscape Architecture and Environmental Design, Finding Center in Theory and Practice*. Baton Rouge: Louisiana State University Press.

Meyer, E. K. (2008). 'Sustaining Beauty. The Performance of Appearance A Manifesto in Three Parts', *Journal of Landscape Architecture*, Spring.

Nehamas, A. (2007). *Only a Promise of Happiness: The Place of Beauty in a World of Art*. Princeton: Princeton University Press.

Nussbaum, M. (1988). 'Nature, Functioning and Capability: Aristotle on Political Distribution', *Oxford Studies in Ancient Philosophy* (Supplementary Volume), 6.

Nussbaum, M. (2000). *Women and Human Development: The Capabilities Approach*. Cambridge: Cambridge University Press.

Nussbaum, M. (2011). *Creating Capabilities*. Cambridge, MA: Harvard University Press.

Sen, A. (1999). *Development as Freedom*. New York: Knopf.

Stern, M. and Seifert, S. (2009). *Civic Engagement and the Arts: Issues of Conceptualization and Measurement*. Washington, DC: Americans for the Arts.

UN-Habitat. (2016). *Habitat III Policy Paper 4 – Urban Governance, Capacity and Institutional Development* (Feb. 29, 2016). [Online]. Available from: http://habitat3.org/wp-content/uploads/PU4-HABITAT-III-POLICY-PAPER.pdf. [Accessed: 17 January 2018].

UN-Habitat. (2017). *Habitat III - The New Urban Agenda* (Oct. 20, 2016). [Online] Available from: http://habitat3.org/wp-content/uploads/NUA-English.pdf. [Accessed: 17 January 2018].

Zuidervaart, L. (2010). *Art in Public: Politics, Economics, and a Democratic Culture.* Cambridge, England: Cambridge University Press.

13 From indicators to face validity to theory – and back again

Measuring outcomes of U.S. creative placemaking projects

Sunil Iyengar

Abstract

Since 2010, the National Endowment for the Arts (NEA) has advanced creative placemaking throughout the United States not only by funding individual projects, but also by providing resources in support of research and evaluation. This chapter reflects on the NEA's journey in promoting metrics and publicly-accessible data sources for assessing practitioners' progress toward achieving livability outcomes—a journey that has led to a theory of change, a logic model, and a measurement model for *Our Town,* the NEA's premier creative placemaking program. These experiences have reinforced the NEA's commitment to understanding and articulating the relationship between *Our Town* program components and the intended and unintended outcomes reported by grantees and their organizational partners. The next phase of this institutional learning process will occur through a mixed-methods evaluation study that will attempt to validate the conceptual framework and metrics that the NEA has developed. Along the way, NEA researchers have grown to appreciate the complexity of concepts and variables at work in empirical studies of creative placemaking. Similarly, NEA researchers have benefited from parallel efforts in the field to identify outcomes and indicators that resonate with policy-makers and practitioners.

Premise

The following is not exactly a personal reminiscence, nor is it a species of case study or policy analysis. What I hope to do, rather, is share an evolving perspective on the research investments that the National Endowment for the Arts (NEA) has made since 2010 to build evidence about creative placemaking – a term that first gained currency from a NEA-commissioned White Paper (Markusen and Gadwa, 2010). The present chapter describes two large research projects that have resulted in metrics for creative placemaking, and it discusses the respective strengths and limitations of both approaches. It is a historical survey, tinged with one arts administrator's discoveries, reappraisals, and, finally, convictions.

Challenges in measuring the impacts of *Our Town*

In the field of program evaluation, it's a commonplace that the best time to start an evaluation is at the start of a program. Certainly, senior managers at the NEA[1] were not slow to recognize that metrics, data analysis, and reporting would be indispensable to any narrative in support of the fledgling *Our Town*[2] program. *Our Town* is the NEA's creative placemaking grant category; its birth coincided roughly with the NEA White Paper, and it preceded (but not by much) the founding of ArtPlace America,[3] a coalition of several national funders, including the NEA. Yet, at the time of *Our Town*'s inception, the NEA's research priorities did not entail a formative or summative evaluation of the program. Instead, the thought was to identify innovative measurement strategies and data sources that could be used to record benefits associated with *Our Town* investments. At no point during this phase did we remotely consider the prospect of drawing causal inferences from NEA funding, where societal or economic benefits might be concerned. Even if we could have performed such daunting methodological tasks as choosing control sites to compare with *Our Town* communities, there would remain problems of scale. How could we assume that the NEA's investment, which often is a relatively low share of a project's overall costs, had had a decisive effect on outcomes typically assessed at the county level, zip code level, or, at best, within a U.S. Census tract?

Enshrining greater livability as a program goal

Before discussing measurement, it is necessary to review the kinds of outcomes that we intended to produce. The *Our Town* program aligned with a key 'objective' in the NEA's previous strategic plan. That objective was 'Livability: American communities are strengthened through the arts' (National Endowment for the Arts, 2014b). The grant application guidelines state:

> The Our Town grant program supports creative placemaking projects that help to transform communities into lively, beautiful, and resilient places – achieving these community goals through strategies that incorporate arts, culture, and/or design ... This funding supports local efforts to enhance quality of life and opportunity for existing residents, increase creative activity, and create or preserve a distinct sense of place ... Our Town requires partnerships between arts organizations and government, other nonprofit organizations, and private entities to achieve livability goals for communities.
>
> (National Endowment for the Arts, 2017b, n.p.)

Having tied *Our Town* to the concept of general 'livability,' the NEA was obligated to define it. The White Paper authored by Markusen and Gadwa (2010) flashed some cues:

> Livability outcomes include heightened public safety, community identity, environmental quality, increased affordable housing and workplace options

for creative workers, more beautiful and reliable transportation choices, and increased collaboration between civic, non-profit, and for-profit partners.

(Markusen and Gadwa, 2010, p.5)

This ambitious rhetoric, implying a need to improve community-level outcomes not historically served by an arts agency of the U.S. government, came at a time when the NEA was pursuing new cross-sectoral and trans-disciplinary initiatives to extend and demonstrate the relevance of its work. Then-chairman of the NEA, Rocco Landesman, had rewritten our motto to 'Art Works,' explicitly recognizing art's functional properties in helping to transform lives and communities. In service of this motto, Landesman and his senior deputy chairman, Joan Shigekawa, encouraged or engineered collaborations with other funders – starting with our much-larger peer agencies within government. Regarding livability and the arts, a prime example was the NEA's work with the U.S. Department of Housing (HUD), the Department of Transportation, and the Environmental Protection Agency to support arts-based projects under the Partnership for Sustainable Communities initiative. This venture gave the NEA unprecedented access to technical knowledge about review criteria and funding decisions affecting community development. But it also allowed the NEA's research team to learn from HUD colleagues about community-level outcomes that they sought to influence through grant-making.

Indeed, there was much learning to be done. Prior to designing a measurement system for tracking the potential impact of *Our Town* grants, the NEA convened a research workshop investigating how to monitor the arts' relationship to the quality of life of communities. Titled *The Arts and Livability: The Road to Better Metrics*, the November 2010 workshop took up some of the trailblazing research by Maria Rosario Jackson, Mark Stern and Susan Seifert, and Stephen Sheppard, and it tapped the expertise of other scholars and consultants, arts leaders, and policy-makers. A report from the convening attests to a plurality of available metrics, but it also signals a raft of decisions – about data sources, analytical methods, and geographic units – that the intrepid researcher must navigate (Pierson et al., 2010). Above all, as should have been apparent to me at the time, one cannot assemble metrics for a creative placemaking program without first investing substantial time and energy in designing a theoretical framework for it.

Birth of NEA Arts and Livability Indicators

Lacking such a framework, we nonetheless saw value in moving swiftly to provide creative placemaking practitioners with research tools of their own. Among NEA management, there was no appetite for a program evaluation *per se*, but rather an understanding that *Our Town* grantees and their partner organizations would welcome guidance on publicly accessible data and introductions to other research resources, so that outcomes associated with creative placemaking could begin to be quantified. Even if we could not test a causal relationship between *Our Town* and greater livability, it seemed worth asking what categories of community-level change were envisioned by our grantees. Beyond this inquiry, we wanted to know if it were possible to construct a series of national indicators, using mainly government data, to chart directions in which *Our Town* communities should be headed, depending on the type of *Our Town* project.

In the end, we led a retrospective review of *Our Town* grant applications – and of applications submitted to a precursor program at the NEA – so that we could sort the desired outcomes of these projects into affinity groups. Fresh from this work, and from an environmental scan, we arrived at four broad dimensions of livability that we connected with *Our Town* programming: residents' attachment to communities; quality of life; local economic conditions; and arts and cultural activity (i.e., the infrastructure supporting artists and arts organizations). We then worked with a consultant to help select indicators from these general outcome areas. The choice of indicators was constrained not only by the outcomes themselves, but also by our requirement that free, publicly accessible data enabling comparisons across U.S. communities be used to populate the indicators. The latter stipulation, in turn, necessitated that the geographic units of measurement were restricted to the zip code area, county, or Census tract – they could go no deeper. Below is the ultimate list of NEA Arts and Livability indicators (Figure 13.1).

Validation exercises get under way

This list resulted from a much more elaborate process of consultation than we had intended. After developing an initial series of candidate indicators, we knew it would be important to gather opinions from grantees and other stakeholders, and to revise the indicators accordingly. Long before the indicators themselves were announced, the NEA had started to field genial criticism about the very approach we were taking. In 2012, Markusen, author of the original NEA White Paper on creative placemaking, posted an online article titled 'Fuzzy Concepts, Proxy Data: Why Indicators Won't Track Creative Placemaking Success.' Her article had been preceded that same year by 'Creative Placemaking Has an Outcomes Problem,' (May 9, 2012) by arts researcher Moss, posted to the same publication. Admittedly, there was then a zeal for indicators – not only on the NEA Research team, but also at ArtPlace America, which had proposed a set of 'vibrancy indicators' to track long-term results of creative placemaking. It's also worth noting that, at the time, HUD and its federal partners had been constructing indicators of their own, as part of an evaluation plan for the Partnership for Sustainable Communities initiative.[4] NEA program managers had been newly inspired to coordinate their efforts with other federal agencies and departments. If we managed to find meaningful metrics for outcomes related to creative placemaking, we reasoned, then some of those metrics could inform other toolkits such as HUD's.

The indicators craze fed partly on the glamor of Big Data, a concept then at its peak within government and the popular consciousness alike. From our

NEA's Arts & Livability Indicators

Outcome Area and the Lowest Geographical Level at which National Data are Available

	Indicator	Geographic level
Resident attachment to community		
C1	Capacity for homeownership (proportion of single-unit structures)	Census tract
C2	Length of residence	Census tract
C3	Proportion of housing units owner-occupied	Census tract
C4	Proportion of housing units occupied	Census tract
C5	Election turnout rate	
C6	Household outflow (tax returns leaving)	
C7	Civic engagement establishments per 1,000 population	Zip code
Quality of life		
Q1	Median commute time	County
Q2	Retail and service establishments per 1,000 population	Zip code
Q3	Violent crime rate	County
Q4	Property crime rate	County
Q5	Percent of residential addresses not collecting mail	County
Q6	Net migration	County
Arts and Cultural Activity		
AC1	Median earnings of residents employed in arts-and-entertainment-related establishments	Census tract
AC2	Proportion of employees working in arts-and-entertainment-related establishments	County
AC3	Relative payroll of arts-and-entertainment-related establishment	County
AC4	Arts, culture, and humanities nonprofits per 1,000 population	Census tract
AC5	Arts-and-entertainment-related establishments per 1,000 population	Zip code
Economic conditions		
E1	Median home purchase loan amounts	Census tract
E2	Median household income	Census tract
E3	Active business addresses	Census tract
E4	Unemployment rate	Census tract
E5	Income diversity	Census tract

Source: *The Validating Arts & Livability Indicators (VALI) Study: Results and Recommendations*, Urban Institute

Figure 13.1 National Endowment for the Arts (2015), The Arts and Livability Indicators.

perspective, the idea was to curate datasets reflecting dimensions of livability that creative placemaking practitioners aspired to affect over time. Critically, the outcome measurements were not meant to communicate the direct impact of creative placemaking projects. Instead, they were intended to permit arts and design organizations and their partners to monitor trends associated with selected outcome areas. Relying on secondary data sources, the indicators also could supplement evaluations of creative placemaking projects. In a November 2012 response to critics of the indicators program, my former colleague Jason Schupbach, then director of Design & Creative Placemaking at the NEA, and I attempted to clarify this key distinction. Beyond recognizing the need to communicate our intent more clearly, the NEA research staff understood that a project of this scope could not afford to rest on the assumption that the indicators we had handpicked in Washington, D.C., would resonate with the breadth of organizations doing creative placemaking work nationwide. In hindsight, the cautionary feedback we had received about the indicators drove us even harder to test their validity.

With the help of the Urban Institute, therefore, we met with *Our Town* grantees and their partner organizations (both on-site and in Washington, D.C.) and showed them maps and tables reporting arts and livability trends in their communities. We also consulted a working group of arts and cultural researchers and evaluators. Among the most impressive features of these 'ground-truthing' sessions were the spontaneous warmth and recognition the citizens exhibited for places on a map and their instant grasp of the values we were attempting to measure. This study, *Validating Arts and Livability Indicators* (VALI), produced a table of indicators and data sources – already shown above – and a detailed report with recommendations. Urban Institute found that 'overall, [respondents] felt the data for the respective indicators for their communities were about right. The data generally reflected respondents' perceptions of local conditions' (National Endowment for the Arts, 2014b). At the same time, respondents would have liked to see data that more closely reflected conditions in the immediate vicinity of the creative placemaking projects. Respondents also expressed mixed views about the relevance of specific indicators to creative placemaking outcomes. Indicators such as 'capacity for homeownership,' 'election turnout,' and 'median commute time' – though justified within the NEA's construct of 'the arts and livability' – did not appeal to some participants. Ultimately, we decided to serve up the entire list of indicators as a menu on the NEA website (National Endowment for the Arts, 2015). In this respect, and on a modest scale, our thinking resembled that of HUD researchers, who finally had opted not to prescribe indicators for their Sustainable Communities grantees, but rather to catalogue hundreds of indicators that could be used to measure progress on 'livability principles' (Partnership for Sustainable Communities, 2014). For our part, we added an online directory of indicators systems and data sources that are not universally available across the U.S., but which can be used by certain communities to track arts and livability-related outcomes. We also posted user case-scenarios illustrating how the national-level indicators can help creative placemaking practitioners. These tools, along with an interactive graphic, achieved about as much public utility for the indicators project as could have expected.

Parallel and complementary research efforts

Even while it pursued national indicators and datasets relevant to creative placemaking, the NEA Research team monitored other studies and measurement techniques that were being used in specific communities nationwide. Much of this work had been directly supported by the NEA – either through its Design division or the Office of Research & Analysis. In Philadelphia, the NEA had awarded funding to help establish *Culture Blocks*,[5] a free, publicly accessible web tool linked to a spatial database of Philadelphia's creative and cultural resources. Similarly, the NEA supported *GEOLOOM*,[6] a cultural mapping collaboration between the Baltimore Neighborhood Indicators Alliance and the Jacob France Institute that proposed to add 'a cultural data dimension that is a vital but oftentimes missing element in the conversations about neighborhoods' (Robert W. Deutsch Foundation, 2017). Beyond providing support for geospatial mapping tools and datasets, the NEA backed *Plan it Hennepin* in Minnesota, a creative placemaking effort that resulted in an evidence-based series of indicators, authored by Metris Arts Consulting. Carl Grodach, Elizabeth Currid-Halkett, Joanna Ganning, Doug Noonan, Mark Treskon, and other researchers worked on NEA-sponsored projects, performing secondary data analyses about the arts and livability. The NEA's research grants program has since expanded to cover primary data collection and experimental or quasi-experimental studies of the arts' benefits for individuals and communities.

Empirical research into variables affecting the livability of communities is a logical function for a national funder of creative placemaking activities. Still, the resulting studies do not necessarily help local policy-makers and artists and designers in their daily efforts to integrate cultural assets with broader community-development projects. Rather than wait until completion of the NEA indicators initiative, the agency's Design team wisely developed other informational resources about creative placemaking and *Our Town* in particular. For example, the NEA's Design team produced a digital storybook titled *Exploring Our Town*, which, to date, features detailed vignettes and key statistics about 78 selected *Our Town* projects (National Endowment for the Arts, 2014b). The team published *How to Do Creative Placemaking*, 'an action guide for making places better' (National Endowment for the Arts, 2016).

Most substantial of all were changes that the NEA made to grant application guidelines for *Our Town*. Specifically, the Design team introduced a knowledge-building grants category to

> expand the capacity of artists and arts organizations to be more effective at executing creative placemaking projects, and to work more effectively with economic and community development practitioners, and vice versa, to improve the livability of the communities and create opportunities for all.
>
> (National Endowment for the Arts, 2017b, n.p.).

For the first time, the NEA solicited grant applications proposing mentorships, training opportunities, technical assistance, and research and policy analysis – and related types of project activity – so that the network of creative placemaking practitioners could be strengthened. As *Our Town* continued to mature, the NEA Design team saw other opportunities to enhance learning that could benefit the creative placemaking field. One such opportunity led the team to another round of collaboration, described below, with the NEA's Office of Research and Analysis.

Building a theory of change and measurement model

In 2016, the NEA Research and Design teams issued a Request for Proposals (RFP) from contractors who could help the NEA to design and conduct a mixed-methods evaluation of *Our Town*. The creative placemaking program then was five years old, and it seemed an opportune time to learn what it had accomplished for individual grantees and for the field at large. The RFP outlined expectations for a 'national survey of all past and current *Our Town* grantees,' a follow-up survey, and seven to ten case studies of *Our Town* communities, all to capture outcomes and processes associated with projects in the NEA's creative placemaking portfolio. There was just one step that needed to be taken before the mixed-methods study could occur. The RFP specified, as the first phase of the evaluation, that the contractor would work with the Design and Research teams to develop a 'theory of change and measurement model' for *Our Town*. This product in turn would stem from a review of NEA administrative data, geographical data covering *Our Town* project activity locations, and insights from stakeholder interviews, a literature review, and 'a rapid scan of logic models, theories of change, outcome measures, and data collection tools for other placemaking initiatives, especially those sponsored by other federal agencies or departments.'

More than a year later, I can admit to having underestimated the amount of time and effort that needed to be devoted to this research project. I had not credited how highly iterative a process would be needed to design a grounded theory for *Our Town*, and to create a measurement model that could account for the variables of interest. After we had hired a contractor to assist us with this crucial first phase, and had been working with this firm for several months, it became clear that bringing in the NEA's Design team would prove vital to interpreting any findings, and that the interim reports from our contractor required extensive vetting. I soon learned that the very same kind of 'ground-truthing' we had identified earlier was necessary now for checking the contractor's assumptions about the data it presented to us. The difference is that this time the validation was being provided not by a cross-section of NEA grantees, but by the program managers for *Our Town*. This approach seemed only fitting, as the primary goal was to design a conceptual framework for the NEA's rationale, processes, and intended outcomes for *Our Town*, so that the agency more effectively could monitor and improve the program's performance.

Before the contract period was over, the NEA's Research and Design teams had refined methods for communicating with each other, and with the contractor, about the agency's changing views of the project, even while keeping within the scope of the contract. Early on, the NEA Research team had encouraged the contractor to think about developing a series of 'nested' logic models that could fit within the overarching theory of change for *Our Town*. We believed that a typology of *Our Town* projects could map onto a typology of *Our Town* outcomes, thus suggesting that each project activity type could generate its own discrete logic model. Instead, we learned from a grants data review that no direct correlation existed between project activity type (e.g., a fair or festival, or an art or design installation) and outcome type (e.g., civic engagement, or beautification of the environment). Accordingly, the contractor's final product consisted not only of a conceptual framework document, with a theory of change and a measurement model, but also a capacious logic model that identified a 'problem statement' for *Our Town*, a program goal, inputs, activities, outputs, and two broad categories of outcome. One outcome is titled 'Innovation/Systems Change and the other is 'Local Community Change.' Each element of the logic model is described meticulously in the framework document itself, which the NEA will make publicly available in 2017/2018. Now that the first phase is completed, the NEA is overseeing a study to validate the conceptual framework (theory of change), logic model, and measurement model (not shown here) for the *Our Town* program. The study will include a national web-based survey of current and former *Our Town* grantees and case studies documenting implementation practices and outcomes achieved through NEA funding (Figures 13.2 and 13.3).

Conclusion

Looking back at the sequence of NEA research investments to understand creative placemaking, one may be forgiven for thinking that we've come full circle. To recap: we began with an indicators system, moved to validation exercises, and then established a theory of change, which itself has bequeathed a new validation study and the quantification of outcomes. From this experience I conclude that future evaluations of interventions operating in complex systems should begin with a sturdy conceptual framework, one co-created by practitioners, researchers, and program managers. I further conclude that national public funders of social-impact work are at their best when they point to real-world examples of how programs look and behave under optimal conditions, and that sometimes the greatest service we can provide is to connect organizations with resources that can be customized for use at the local level. For this dual reason – to honor the need for an organic theory behind each program or intervention, but also to promote public access to data and tools that can be adapted ingeniously to suit the context – I have no difficulty justifying the two distinct learning pathways (one an indicators system, and the other a conceptual framework) that have informed the NEA's understanding of the *Our Town* program and its associated outcomes.

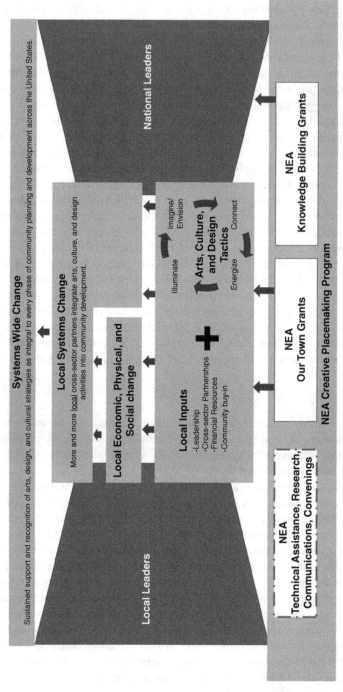

Problem Statement: American communities everywhere face a distinctive set of local economic, physical, and/or social challenges. And yet, community leaders are often unaware of solutions that can arise from the successful adoption and integration of arts, design, and cultural strategies.

Figure 13.2 National Endowment for the Arts (2017b), *Our Town's* Theory of Change.

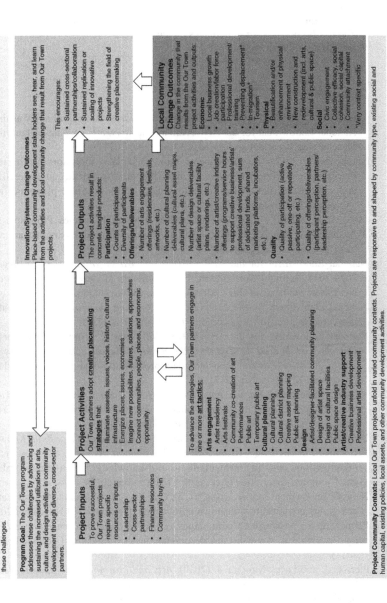

Figure 13.3 National Endowment for the Arts (2017b), *Our Town's* Logic Model.

Notes

1 National Endowment for the Arts, www.arts.gov/
2 *Our Town*, www.arts.gov/grants-organizations/our-town/introduction
3 ArtPlace America, www.artplaceamerica.org/
4 Partnership for Sustainable Communities, www.sustainablecommunities.gov/
5 *Culture Blocks*, www.cultureblocks.com/wordpress/
6 *GEOLOOM*, https://geoloom.org

References

Markusen, A. and Gadwa, A. (2010). *Creative Placemaking*. [online] Washington, D.C.: National Endowment for the Arts. Available at: www.arts.gov/sites/default/files/Creative Placemaking-Paper.pdf [Accessed: 13 November 2017].

Markusen, A. (2012). 'Fuzzy Concepts, Proxy Data: Why Indicators Won't Track Creative Placemaking Success', *Createquity*. Available at: http://createquity.com/2012/11/fuzzy-concepts-proxy-data-why-indicators-wont-track-creative-placemaking-success/ [Accessed: 13 November 2017].

Moss, I. D. (2012). 'Creative Placemaking Has an Outcomes Problem', *Createquity*. Available at: http://createquity.com/2012/05/creative-placemaking-has-an-outcomes-pro blem/ [Accessed: 13 November 2017].

National Endowment for the Arts. (2014a). *Art Works for America: Strategic Plan*, FY 2014-2018. [online] Washington, D.C. Available at: www.arts.gov/sites/default/files/NEAStrategicPlan2014-2018.pdf. [Accessed: 13 November 2017].

National Endowment for the Arts. (2014b). *Exploring Our Town*. Available at: www.arts.gov/exploring-our-town/ [Accessed: 13 November 2017].

National Endowment for the Arts. (2015). *NEA Arts & Livability Indicators: Assessing Outcomes of Interest to Creative Placemaking Projects*. Available at: www.arts.gov/artistic-fields/research-analysis/arts-data-profiles/arts-data-profile-8/arts-data-profile-8 [Accessed: 13 November 2017].

National Endowment for the Arts. (2016). *How to Do Creative Placemaking*. [online] Washington, DC Available at: www.arts.gov/sites/default/files/How-to-do-Creative-Place making_Jan2017.pdf [Accessed: 13 November 2017].

National Endowment for the Arts. (2017a). *Our Town: Projects that Build Knowledge about Creative Placemaking – Grant Program Description*. Available at: www.arts.gov/grants-organizations/our-town/projects-that-build-knowledge-about-creative-placemaking-grant-program-description [Accessed: 13 November 2017].

National Endowment for the Arts. (2017b). *Our Town: Introduction*. Available at: www.arts.gov/grants-organizations/our-town/introduction. [Accessed: 13 November 2017].

Partnership for Sustainable Communities. (2014). *Sustainable Community Indicators Catalog*. Available at: www.sustainablecommunities.gov/indicators [Accessed: 13 November 2017].

Pierson, J. and Lacey-Moreira, J. (2010). *Arts and Livability: The Road to Better Metrics*. [online]. Washington, D.C. Available at: www.arts.gov/sites/default/files/Arts-and-Liva bility-Whitepaper.pdf [Accessed: 13 November 2017].

The Robert W. Deutsch Foundation. (9 March 2017). GEOLOOM Launch Planned for Summer 2017. [online]. Baltimore, MD. Available at: https://rwdfoundation.word press.com/2017/03/09/geoloom-launch-planned-for-summer-2017/ [Accessed: 1 June 2018].

Schupbach, J. and Iyengar, S. (2012). 'Our View of Creative Placemaking, Two Years In', *Createquity.* Available at: http://createquity.com/2012/11/our-view-of-creative-placemak ing-two-years-in/ [Accessed 13 November 2017].

Urban Institute. (2014). *The Validating Arts & Livability (VALI) Study: Results and Recommendations.* [Online]. Washington, D.C: National Endowment for the Arts. Available at: www.arts.gov/sites/default/files/VALI-Report.pdf [Accessed: 13 November 2017].

Conclusion

Moving into the beyond – What's next for creative placemaking?

Anita McKeown and Cara Courage

As a nascent field, nearing completion of its first decade, creative placemaking is still evolving. As the concept is embraced and enacted through practice globally, this book has sought to present approaches and practices that have moved away from the popularly attributed elements of creative placemaking – of monumental artworks and artists' live-work spaces and the like – to more diverse practices and voices that both offer sector critique and development. This concluding chapter draws attention to the underlying concerns that motivated the book and the questions and issues raised within it, signalling the 'move beyond' the current into creative placemaking's next era.

Part 1 – Coming full circle

The aim for the book was never to be a definitive text on creative placemaking; instead, the motivation was to present and integrate a practitioner-led critique in conversation with an academic and research-based discourse. Many of the authors, and ourselves as editors, traverse research and practice and value the different forms of mutually-informing knowledges this traversal produces. In this sense, our professional experience within ephemeral, situated and citizen-led creative practices guided the initial Call for Papers and the curation of the conference sessions (see Preface and Introduction). Our initial conversation presented discussions and practices that we recognised from non-object-orientated art practices that were becoming increasingly relevant within creative placemaking's emergent practice and are now offered as a springboard for the next phase of the sector.

Many of this volume's contributors are anchored in creative practices that are often process-driven and that move towards a complex and more nuanced practice and understanding of creative placemaking. They present tangible explorations and responses to many of the concerns that have been prevalent within creative placemaking discourse, such as the role of the artist and the social, economic and political contexts of this work. We are not alone: these concerns have also fuelled common discussions within non-object-orientated art practices for at least half a century. From this perspective the book does not seek to provide answers but rather charts a dynamic conversation through the collection of professional creatives' responses to key commentaries prevalent since creative placemaking's inception. The practitioners presented here have a long-standing engagement with 'arts in place' practices and as such their contributions offer a comprehensive survey and an insightful critique for the creative placemaking scholar. Thus, this volume is presented as a reflection on what creative placemaking's evolution could mean for future practitioners and scholars within the context of emergent global challenges and radical changes to where and how places are made and sustained.

Despite many of the criticisms levelled at creative placemaking, the book is an optimistic and positive consideration of creative placemaking's evolution and its potential. Emerging from a policy leadership initiative in the USA, local and global uptake of creative placemaking has been informed by increasing involvement from diverse arts practitioners and organisations. This collection, therefore charts a departure from economic impact towards broader intrinsic values of the arts and citizen engagement and ephemeral practices of creative placemaking. By drawing attention to the ongoing concerns around diversity, displacement and concepts of belonging such engagement has informed a positive move towards participation in the arts contributing towards Bedoya's (2013) demand for 'place-keeping' and Pritchard's (this volume) call for 'place-guarding'. Issues of place attachment and belonging, present in discussions of creative placemaking from early in its inception, highlight concerns around 'who, how and why we should support arts and cultural activity' (Markusen and Gadwa, this volume) contribute to the book's overarching question: how will creative placemaking evolve over the next decade and beyond?

Many of the projects that were included within Markusen and Gadwa's (2010) White Paper illustrate aspects of what could be considered as leading through practice, an aspect often overlooked when engaging with the field of creative placemaking: the projects presented drawn from the field were motivated not by economic agendas but an encouragement of inclusive, participatory and collaborative processes beyond the limits of elitist art practices and the sanctified walls of the gallery (and indeed, a similar move to generative planning from a professionalised silo (Courage, 2017)). However, this was not necessarily reflected in many of the early projects funded by the National Endowment for the Arts (NEA)[1] and ArtPlace,[2] which tended to conform to capital spend projects, design enhancements, festivals or monumental artworks. This can be seen to have perhaps overshadowed the wider, and perhaps more grassroots, culturally creative foundation and potential of creative placemaking and contributed to many of the concerns around elitism and exclusion and critiques of its practice.

The book argues for how creative placemaking – defined in its broadest sense as creative practices within the process of placemaking – can offer avenues for understanding, galvanising and expressing place identity and place attachment through individual, cultural, social and material articulations of psychosocial processes (Courage, 2017). The material process of creative placemaking involves social and cultural processes that can uncover senses of self with the community and in place, considerations of disbelonging (Bedoya, 2013), and inclusion and exclusion in place habitation and politics. Opportunities for citizens to engage and co-create within creative placemaking are necessary if the sector is not to be complicit in nor perpetuate social displacement but instead contribute to an authentic and meaningful sense of place, and a sense of ownership of and belonging. The ethical imperatives of social and environmental justice that should be necessarily embodied in a concept such as creative placemaking are manifest in concepts of identity and implemented through an articulation of not only an understanding of place, but also having a place within it. The concerns around the addition of the exclusivity of the 'creative' modifier for placemaking (Bedoya, 2013; Kent, 2013; Mehta, 2012), including the very real psychological barriers to involvement in the arts (Keaney, 2008; McCarthy and Jinnett, 2001; Sinnott, 1983) was a key motivation behind the curation of the conference sessions, in turn informing this volume. Our professional experience sought to introduce the fifty-year heritage of non-object-orientated practices (Lippard, 1968) and place-based creative practices that provide a heritage that is often overlooked within creative placemaking (Courage, 2017). Such practices, for example, from land art to New Genre Public Art (Lacy, 2008), have consciously explored complexity and creative self-expression, and interrogated the language and meanings of community, partnership, identity and dynamics of power and spatial equity. This book presents practitioner-led expansion of the potential of creative placemaking, drawing on this established legacy and integrating the arts, creativity and the role and value of creative knowledge and skills that is found within suburbs, villages, rural areas and cyberspace (Edensor and Millington, this volume). Through their engagement with the social ecosystems and infrastructures that are engaged in creative placemaking, our contributors offer a critical engagement with and inclusion of the arts and creativity within civic life beyond the expressive pursuits of a culturally dominant elite. Their contributions challenge many of the creative practices, historically utilised within previous iterations of cultural regeneration, and as such the book offers an understanding of creative placemaking that forefronts creative practices that are situated, participatory and collaborative in nature. Importantly for creative placemaking, the ecologies of practices the book presents form a collective challenge to the limitations and dangers of harnessing creativity's intrinsic values in the pursuit of geopolitical economic agendas. Co-opting these values into the creative industries for economic gain or imposing reductive elitist understanding of creativity can have profound effects. Yet creative practices, when unconstrained by disciplinary silos or economic motivations, can offer expanded opportunities to engage in new conversations and creative social exchange. The following section explores these issues in more depth.

Part 2 – The art of creative placemaking

Artists and arts have a history of being employed as 'agents of change' (Stern and Seifert, 2006; Matarasso, 1997; Miles, 2000). Yet this should not be confused with a vision of arts as a way of 'remaking communities, creating jobs, reducing crime, in short saving the world, in light of long-standing systemic inequities' (McKeown, 2015a). Lippard's (1968) seminal text, *Six years: The De-materialisation of the art object*, and Kraus' (1979) *Sculpture in an expanded Field* are early documents that began to highlight the concerns around equity, power dynamics, participation and authorship that arose with non-studio art practices. More contemporary texts, such as Kwon's *The Wrong Place* (2000) or McGonagale's work on reciprocity, negotiation and active citizenship (2011, 2010, 2007) engage in a critical stance as practitioners working in collaborative, participatory or activist practices. Kwon's (2004, p.60) term 'art in the public interest' identifies the engagement with 'social issues, political activism and community collaborations'. Such practices do not take place within the gallery but instead often work across a range of non-art contexts with numerous constituents. By rejecting the classical notion of the creative genius (McGonagle, 2010) post-studio practices offer an opportunity to explore and encourage everyday creativity within creative place-making. These practices enable new relationships to form and facilitate 'the constant imagination, search for, and construction of alternatives' (Hays, in Osman et al., 2002, p.58). Thompson understands such practices as tactical forms of resistance to the current global context of an 'increasingly privatised and visualised cultural sphere' (Sholette and Thompson, 2004, p.52).

Within the context of creative placemaking, these practices can be considered as transgressions into or subversions of the functions and power mechanisms that construct dominant place narratives, a key concern when considering belonging and displacement. Creative placemaking projects can have an interstitial function as 'third spaces' (Soja, 1996) where, as grounded in an inclusive systemic approach, there is opportunity to explore and reconsider narrowly understood or ill-defined place-based situations and harness its potential to facilitate understanding and scalable localised solutions to systemic problems. Many of the projects presented here develop micro 'network[s] of resistance' (Sholette and Thompson, 2004, p.52) as a bi-product of their creation, connecting citizens and organisations through their expanded field of operation and shared aims. As an inclusive approach, this offers a way into generative spatial planning, critical decision-making and policy development for those often excluded. By recognising and valuing multiple actors and their creativity within creative placemaking, the projects can begin to transcend polarising debates of expert/non-expert and offer alternative paths to the officially sanctioned channels that normatively create the places we live.

As diverse practitioners are increasingly recognised in and engage with creative placemaking, the sector begins to reflect contemporary approaches to creativity that challenge or do not conform to an understanding of the arts and creative processes often held by those outside the realms of contemporary creative practices. Increasingly, those charged with the commissioning and engagement of creative practitioners within placemaking contexts are acknowledging more nuanced understandings and manifestations of professional creative practice. Currently, developments within the NEA and ArtPlace's approach to funding projects are reflective of the move away from cultural corridors, live/work schemes and monumental practices, for instance. Furthermore, the knowledge from creative practitioners' and ephemeral processual practices of creative placemaking are being formally recognised and integrated into creative placemaking, as evidenced by the NEA's two-day symposium *Beyond the Building* (NEA, 2014) and MIT's *Places in the Making* (Silberberg, 2013) report. *Beyond the Building* discussed the importance and potential of ephemeral art practices, while MIT's report identified the potential of processual and continual placemaking, stressing the need for an open source approach.

However, it should not be forgotten that ephemeral and processual forms of creative placemaking can also have negative aspects, which need to be considered. Contested place narratives are at the core of creative placemaking practices and can be exacerbated by ill-considered interventions and the policy and funding behind them. Temporary place activation programmes are often inadequate at nurturing inclusive practices that sensitively address the relationship between 'place, affect and memory' (Barns, this volume). Questions of what should and could be celebrated and by whom, when explored within public culture, 'can often be highly fractured and discordant' (ibid.). The authors presented in this book all engage with multi-form presentations and ephemeral practices which can offer an intimate means of experiencing the multiplicity of narratives that shape a project's location. Yet they are also realistic in recognising the limitations of such projects. Temporary projects cannot and should not be expected to resolve contested narratives that have been embedded over generations. If undertaken sensitively they can 'enable multiple forms of attachment to be acknowledged, while also allowing new connections to be forged' (ibid.). The growing awareness of temporary projects as a critical first step in community engagement as part of place activation is a global trend, enshrined in Reynold's (in Project for Public Spaces, 2013) 2009 maxim, 'Lighter, Quicker, Cheaper' (LQC) and championed by numerous other placemaking agencies since. Any benefits will only be apparent if undertaken sensitively and with an awareness of the potential impacts of any resulting consequences (McKeown, 2015b) and not deployed due to an appeal to any perceived low-cost or quick-fix political and material end-solution finding.

Like any intervention, a temporary action can have negative and positive outcomes. This important aspect is illustrated through Urban Vision's consideration of LQC (Pathak, this volume) within three localities in Mumbai, India, and Alpalhão's (this volume) reflexive critique of participatory practice, offering an insight into the considerations and consequences of using interventional practices developed within different contexts. Urban Vision's exploration of the creation of human-scaled social space within a rapidly-changing urban context with a history of social space as integral reaffirmed the need to understand local contexts and adapt practices accordingly. Meanwhile, Alpalhão's chapter offers a consideration of the role of the artist and public apathy and disinterest in place, stemming from feelings of being excluded from its decision-making and planning. These issues are discussed throughout this book, of course, with the chapters by Edensor and Millington, Pritchard, and Walker and Marsh particularly pertinent.

The changing context in Mumbai has raised issues that many western contexts have already experienced: an auto- and infrastructure-centric environment is being created or maintained that is to the detriment of social spaces. Such practices are useful for their opportunities to explore possibilities initially – as Lydon and Garcia (2015) rightly attest to Tactical Urbanism – but they require careful consideration and sophisticated understanding of context and complex skills if they are to be beneficial and facilitate inclusive practices. This has import also for the global uptake of creative placemaking, with many countries still struggling to address equitable placemaking practices. As such, creative placemaking interventions become deeply political acts that contribute to defining and constituting place. To this end creative placemaking practitioners should engage with a full life-cycle analysis of their proposed interventions and awareness of the complexities of the social construction of place. Further, practitioners must beware of the potential for creative placemaking to be appropriated into the use of celebrating place identity and unique culture as a way to 're-image and sell places' (Ward, 1998. p.193). Counterintuitively perhaps, creative placemaking can offer alternatives to the attributes of placemaking and branding that are increasingly attributed to it by not being 'tethered to a meaning of place manifest in the built environment' (Bedoya, 2013) through the creation of spaces that encourage multiple, new or different understandings of place. However, this takes skill and a critical understanding of the power and historical dynamics at play.

This book introduces an array of contemporary creative professional practices and approaches, illustrating the diverse skillset necessary for interdisciplinary practices and complex contexts that engage with issues of social and environmental justice. It is in these new territories of practice, that the book's authors re-imagine traditional roles with additional skills such as an understanding of positioning and power dynamics, an ability to negotiate and the acknowledgement that it can no longer be 'business as usual' when it comes to authority. Their projects reflect the ecology of knowledge and expertise embedded in processes that emerge from the inclusion of a broad range of constituents, requiring trust, communication skills, sharing information, a willingness to learn and adapt, all qualities of a more complex systemic approach. This requires dynamic leadership models that demand specific expertise at different times while offering foundations for multiple opportunities to participate which have expanded the concept of stakeholding. Leadership, coming from a creative catalyst, has been a key element in creative placemaking from its inception and the book's contributors begin to expand the understanding of this aspect of creative placemaking, in new ways. Increasingly, research on leading-through-artistic-practice (Frye-Burnham, 2007) is now being incorporated into emerging ideas on leadership and consequently into creative placemaking practices.

Each chapter in this book illustrates this changing landscape by sharing the multiple methods that can encourage new ways of seeing the world beyond the concept of the professional creative as sole creator. In this instance, the practitioners, through multiple partnerships, form ecologies of practice to engage with specific knowledge that subsidised their own skillset, while integrating the benefits of their field-specific knowledge. Reconsideration of the role of the professional creative, in conversation with active citizens and other disciplines, illustrates how such skillsets can be used to create sites for intersectional issues and multiple identities to be enacted. McKeown (2015a) has identified thirteen distinct roles attributed to artists, highlighting a range of skills of value for creative placemaking, *Collaborator*, as *Negotiator*, as *Mediator* and *Catalyst*. These diverse roles reflect the multiple modes of operation that many of the book's practitioners currently employ and are increasingly expected to employ within their professional practice. Overtime, this approach develops a transfer-able interdisciplinary skillset with a range of constituents within various contexts.

This understanding of contemporary creative practices will do much to address the concerns of the 'creative' modifier of creative placemaking and the critique of creative placemaking's potential to be exclusive. Indeed, within the new socio-economic and environmental conditions predicted, the continued evolution of reciprocal models of practice to re-connect 'art's ethical responsibilities with its aesthetic responsibilities' (McGonagle, 2011, p.44) becomes a vital task for creative placemaking. The book's authors collectively contribute to an expanded understanding of professional and intersectional creative working practices that that can be traced back to earlier practitioners such as Beuys (see *Social Sculpture*, 1972),[3] Artist Placement Group (1969)[4] and Mierle Laderman Ukeles (see *Maintenance Art*, 1969).[5] Although the heritage of a 'participatory, socially interactive framework for art' (Gablik, 1991, p.7) is now common practice for many artists, this approach has not yet been fully integrated into creative placemaking. Acting as 'mediators, creative thinkers or agitators' (Bourriaud, in Doherty, 2004, p.10), this is increasingly recognised within contemporary crea-tive placemakers such as Candy Chang,[6] Theaster Gates[7] and the Rebuild Foundation,[8] as well as Rick Lowe and Project Row Houses.[9]

Part 3 – Into the beyond

This volume's closing chapter highlights the need for trial and error, exploratory approaches, listening to multiple voices and experiences and being open to change, a common thread that runs through the book. As the USA sector heavyweights, the NEA, reaches a pivotal stage in its engagement with the concept of creative placemaking and ArtPlace completes its deep dive evaluative Field Scans[10] before its operational closing in 2020, there is an evident, at this macro level, a move away from the historic valuations of creative placemaking. The shift from a 'hyperfocus on economic advocacy' (Markusen and Gadwa, this volume) towards more holistic and expansive notions of value and success is a timely milestone. Will recent changes to the USA's administration dash this hope or will creative placemaking continue to morph and evolve without philanthropic support given the global spread and take-up of the concept? Indeed, creative placemaking's journey to date draws attention to the intrinsic qualities that the editors perceive in its potential as a practice: a dynamic evolving approach and set of tools that will continue to be enacted globally by many. Whether it is called creative placemaking in the future (though this term has evident sticking power and understandably pragmatically so) and operates as a strategic agenda is a different matter.

Rather than a neat conclusion, the book leaves us with a number of critical questions. To list but some: will the value of creativity in making and re-making our world have become more centralised or will the arts have been abandoned, side-lined or ever more tokenised as geopolitical agendas focus more on technology and environmental challenges? Or will the current interest in social design thinking and the benefits arts and humanities training brings be deemed critical to our continued existence? What should and could be the role of the developer, as increasingly, the only constituent around the planning table with any funds for creative place-based endeavour, be? And how will theory and practices interplay and be evaluated, curated, disseminated, collected and exhibited?

As creative catalysts, many practitioners are challenging their fields' conventional practices of gallery presentation of objects and the design and construction of buildings. By moving beyond normative roles of practice the perspectives from these practitioners offer insights into a critical reflection of standard practices, thus informing both the field of creative placemaking and the education and training of those working professionally in disciplines that are engaged in the sector, from housing and education through to transit and health. All the authors in the book are expanding beyond their fields and the skills that they require are not yet integrated into the education and training of emerging practitioners. As a nascent field, creative placemaking-specific courses are few in number and mentoring opportunities for younger practitioners are limited, not least as established practitioners are in the process of 'learning as they go'. Granted there are many examples of best practice from other socially engaged disciplines and earlier theorists but the planning schools, art colleges and architectural programmes are only beginning to embrace the need for skills beyond formal language, building and planning regulations, such as advocacy skills, cultural awareness and community liaison, and not least, the disposition of the creative practitioner to work in relative expertism, not as sole author (Courage, 2017). Creative placemaking is manifest in communities as a 'system of interrelated elements' (Davidoff, 1965, p.337) and organisms. This perspective inevitably necessitates an even greater understanding and implementation of an ecological approach than may have been initially recognised, when included in Markusen and Gadwa's White Paper and SIAP's *Ecology of Cultural Revitalisation* (Stern and Seifert, 2006). Other fields are also embracing a systemic approach as the realisation of the impact of isolated practices and knowledge comes to the fore and existential threats from a range of source make themselves known. Many practitioners in a broad range of fields are looking to create partnerships that could be said to manifest what Gablik (1992, p.22) called a 'radical relatedness'. The diverse skills of the artist and other's expertise within a creative experimental approach can form Gablik's (ibid., p.2) 'connective aesthetic', a means to hold the ecological diversity inherent in place in ethical and equitable ways. From this perspective perhaps the global push and engagement with the Sustainable Development Goals (2015) could offer a common ethos to underpin creative placemaking?

In this way, the editors understand creative placemaking to be centrally positioned to deal with the challenges of the twenty-first century with creative place-based approaches offering tangible locally scaled explorations of global concerns. The original creative placemaking White Paper was initially presented to the Mayors' Institute on City Design (MICD), a leadership initiative of the NEA in partnership with the United States Conference of Mayors (USCM) and American Architectural Foundation (AAF). MICD seeks to 'prepare' mayors to 'be the chief urban designers of their cities' (MICD, 2015). As professional creatives working within the public realm no longer see themselves as sole creative genius, the threat of human-driven climate change and the resulting problems becoming increasingly acknowledged, it is clear that all citizens will need to contribute to placemaking practices. No longer can mayors and other professionals engaged in spatial planning be considered the 'chief urban designers of their cities'.

It is noteworthy that an analysis of creative placemaking definitions by twelve key actors in the field identified no overt statement of environmental criteria (McKeown, 2015a). Furthermore, all the chapters within this book consider human-centric approaches, highlighting one of creative placemaking's pressing vulnerabilities – its human-centric default. The editors offer that a beyond-human lens is severely lacking from the creative placemaking field, yet this addition offers an important direction for future research if creative placemaking as a sector is to contribute to the development of sustainable and resilient places. Other scholarly work has addressed the need for the integration of nature and culture (Weintraub, 2012; Wright, 2005). Moving towards a bio-psychosocial approach (McKeown, 2015a), which would expand concepts of belonging, self and other to include other organisms and processes that recognise their contribution to the practice of placemaking. As an ethical imperative, this integrates the social and environmental justice that the editors understand is the core of creative placemaking, beyond economic motivations. Although creative placemaking was informed by a cultural ecology approach to revitalisation (Stern and Seifert, 2006), the next stage in its evolution will need to expand the understanding of what an ecological approach truly means. The duty to ensure the 'identification, protection conservation, presentation and transmission to future generations of the cultural and natural heritage' is embedded in UNESCO's World Heritage constitution, article 4 (1973). The recognition of the importance of both cultural and natural heritage for future generations offers a useful underpinning in this vein as well as an ethical imperative to expand creative placemaking's understanding of an ecological approach.

Returning to non-studio and expanded creative practices, many arts practitioners are no longer working only with self-expression, instead applying their skills to 'wicked' problems 'in pursuit of a sustainable planet' (Weintraub, 2012). The inclusion of STEAM (Science, Technology, Engineering, Arts and Maths)

(Maeda, 2010) educational practices into creative placemaking has much to offer the field. As an integration of arts and science, the STEAM project-based pedagogical approach could encourage younger generations to engage with situated learning and global systemic concerns through local contexts. Place-based learning provides opportunities to embed a student's local context and personal experience as a foundation for their own learning and creative exploration. As a tangible process, STEAM education offers an opportunity for students to embed and apply their knowledge in new ways. Moving between tacit, explicit and abstract knowledge could enable a sophisticated response to place concerns, with a deeper understanding of place-systems leading to an integration of social design thinking. This approach incorporates the creative processes and strategies common to artists and designers into a resolution of issues within social contexts. Creative placemaking as a policy platform will have to address existential concerns if it is to remain relevant, requiring systemic knowledge and a social and environmental ethos deeply embedded into its practices. However, as the book has identified, through this initial tracking of creative placemaking's journey, the inclusion of creative practitioners must be integrated from the beginning to ensure that the value and benefits from the field and nuanced understanding of practices and processes are present. Just like creative placemaking's first decade, without a nuanced understanding of STEAM, we could see the benefits and potential of STEAM, albeit compromised by economic agendas and lack of awareness of its heritage, practices and evolution.

At the present time, there are many culture-led organisations and practitioners that have established best practices and are continuing to contribute to multiple forms of conversations with others involved in making places creatively – what will creative placemaking, and its place, look like in another ten years' time? This volume identifies rich seams for future exploration that would be necessary to sustain the vitality and potential inherent in the combination of words 'creative placemaking' into its next decade and beyond. Together, all contributors have, in their practices and in these chapters, worked to break down the assumptions of creative placemaking and build it anew from a global and multi-layered purview and sitting across urban design and art heritage, theory and practice, planning and policy and cultural and place heritage and politics. At the beginning of this book we asserted that creative placemaking is reductively considered a subset of placemaking and valued for its economic instrumentalisation of the arts. As we have since seen, creative placemaking is much more interesting, complex, vital and exciting than that.

Notes

1 National Endowment for the Arts, www.arts.gov
2 ArtPlace America, www.artplaceamerica.org
3 Beuys (1972), *Social Sculpture*, www.tate.org.uk/art/art-terms/s/social-sculpture
4 Artist Placement Group, www2.tate.org.uk/artistplacementgroup/
5 Mierle Laderman Ukeles (1969), *Maintenance Art*, www.arnolfini.org.uk/whatson/mierle-laderman-ukeles-maintenance-art-works-196920131980

6 Candy Chang, http://candychang.com
7 Theaster Gates, www.theastergates.com
8 Rebuild Foundation, https://rebuild-foundation.org
9 Project Row Houses, https://projectrowhouses.org
10 ArtPlace America Field Scans, www.artplaceamerica.org/our-work/research/translat ing-outcomes

References

Alpalhão, L. (2018). '*Outros Espaços*: Apathy and lack of engagement in participatory processes', this volume.

Barns, S. (2018). 'Arrivals and departures: Navigating an emotional landscape of belonging and displacement at Barangaroo in Sydney, Australia', this volume.

Bedoya, R. (2013). *Creative Placemaking and the politics of belonging and disbelonging.* Available at: www.worldpolicy.org/blog/2013/05/13/creative-placemaking-and-politics-belonging-and-dis-belonging. [Accessed: 25 March 2018].

Beuys, J. (1972). Social Sculpture [online]. Available at: www.tate.org.uk/art/art-terms/s/social-sculpture [Accessed 30 May 2018].

Courage, C. (2017). *Arts in Place: The Arts, the Urban and Social Practice.* Abingdon: Routledge.

Davidoff, P. (1965). 'Advocacy and Pluralism in Planning', *Journal of the American Institute of Planners*, Vol. 31, no. 4.

Doherty, C. (2004). 'The new situationists' in *Contemporary Art: From Studio to Situations.* London: Black Dog Publishing Ltd.

Edensor, T. and Millington, S. (2018). 'Spaces of vernacular creativity reconsidered', this volume.

Frye-Burnham, L. (2007). 'The artist as leader' in Douglas, A. and Fewmantle, C. (eds.,) *Leading Through Practice.* Newcastle: AN Commissions.

Gablik, S. (1991). *The Re-Enchantment of Art.* London: Thames and Hudson Press.

Gablik, S. (1992). 'Connective aesthetics', *American Art*, Vol. 6, no. 2 (Spring), pp. 2–7. Available www.jstor.org/stable/3109088 [Accessed: 25 March 2018].

Hays, K. M. and Kogod, L. (2002). 'Twenty projects at the boundaries of the architectural discipline examined in relation to the historical and contemporary debates over autonomy', *Perspecta* 33, pp. 54–71. Available from: www.jstor.org/stable/1567297. [Accessed: 25 March 2018].

Keaney, E. (2008). *From Indifference To Enthusiasm: Patterns Of Arts Attendance In England.* London: Arts Council England.

Kent, F. (2013). *All Placemaking Is Creative: How A Shared Focus On Place Builds Vibrant Destinations.* Available at: www.pps.org/reference/placemaking-as-community-creativity-how-a-shared-focus-on-place-builds-vibrant-destinations/. [Accessed: 25 March 2018].

Krauss, R. (1979). 'Sculpture in an Expanded Field', *October*, Vol. 8, no. Spring, pp. 30–44.

Kwon, M. (2000). 'The wrong place', *Art Journal* Vol. 59, no. 1 (Spring), pp. 33–44.

Kwon, M. (2004), One Place after Another, Cambridge, MA: MIT Press.

Lacy, S. (2008). 'Time in place: New genre public art a decade later' in Cartiere, C. and Willis, S. (eds.,) *The Practice of Public Art.* New York City: Routledge.

Laderman-Ukeles, M. (1969). Manifesto for Maintenance Art [online]. Available at: www.arnolfini.org.uk/blog/manifesto-for-maintenance-art-1969 [Accessed 30 May 2018].

Lippard, L. R. (1968). *Six Years: The Dematerialisation of the Art Object 1966 – 1972.* Berkley: University of California Press.

Lowe, R. (2013). *Project Row Houses at 20*, Creative Time Reports 7, Oct 2013. Available at: http://creativetimereports.org/2013/10/07/rick-lowe-project-row-houses/. [Accessed: 20 December 2017].

Lydon, M. and Garcia, A. (2105). *Tactical Urbanism*. Washington: Island Press.

Maeda, J. l. T. (2010). 'Innovation is born when art meets science', *The Guardian* [online]. Available at: www.theguardian.com/technology/2010/nov/14/my-bright-idea-john-maeda [Accessed: 11 April 2015].

Markusen, A. and Gadwa, A. (2010). *Creative Placemaking*. Washington, DC: Mayors' Institute on City Design and the National Endowment for the Arts. October. Available from: www.arts.gov/publications/creative-placemaking [Accessed: 23 December 2017].

Markusen, A. and Gadwa Nicodemus, A. (2018). 'Creative placemaking: Reflections on a 21st-century American arts policy initiative' this volume.

Matarasso, F. (1997). *Use or Ornament: The Social Impact of Participation in the Arts*. Stroud: Comedia.

Mayors Institute of City Design (MICD). (2015). What is the Mayor's Institute? [online]. Available at: www.micd.org/about/ [Accessed: 15 January 2013].

McCarthy, K. F. and Jinnett, K. (2001). *A New Framework for Building Participation in the Arts*. Santa Monica, CA: RAND Corporation. Available at: www.rand.org/pubs/mono graph_reports/MR1323. [Accessed: 25 March 2018].

McGonagle, D. (2007). 'Forward' in Butler, D. and Reiss, V. (eds.,) *Art of Negotiation*. Manchester: Cornerhouse.

McGonagle, D. (2010). *Passive to Active Citizenship: A Role for the Arts*. Conference paper, *Bologna in Context*, Dublin. 24 October. The Honorable Society of King's Inns, Dublin.

McGonagle, D. (2011). 'An "Other" Proposition - Situating Reciprocal Practice' in Parry B. (ed.) with contributors Medlyn, S. and Tahir, M., *Cultural Hijack: Rethinking Intervention*. Liverpool University Press, pp. 40–48.

McKeown, A., (2015a). 'Cultivating permaCulural resilience; towards a creative place-making critical praxis'. Unpublished PhD thesis, National College of Art and Design Dublin.

McKeown, A. (2015b). 'Deeper, Slower, Richer: A slow intervention towards resilient places in placemaking', *Edge Condition*, Vol. 5 (January). Available at: www.edgecondi tion.net/uploads/9/5/5/6/9556752/edgecondition_vol5_placemaking_jan15.pdf. [Accessed: 25 March 2018].

Mehta, N. (2012). *The Question all Creative Placemakers should Ask*. Available at: http:// nextcity.org/daily/entry/the-question-all-creative-placemakers-should-ask. [Accessed: 14 June 2013].

Miles, M.F.R. (2000). 'Art and social transformation – theories and practices in contemporary art for radical social change'. PhD. London: Chelsea College of Art.

National Endowment for the Arts. (2014). *Beyond the Building, Performing Arts and Transforming Place*. Washington, DC: National Endowment for the Arts.

Pathak, A. N. (2018). 'A case for human-scale social space in Mumbai', this volume.

Pritchard, S. (2018). 'Place guarding: Activist art against gentrification', this volume.

Project for Public Spaces. (2013). *The Lighter Quicker Cheaper Transformation of Public Spaces*. Available at: www.pps.org/reference/lighter-quicker-cheaper/. [Accessed: 18 January 2013].

Sholette, G. and Thompson, N. (eds.). (2004) *The Interventionists: Users' Manual for the Creative Disruption of Everyday Life*. Cambridge, MA: MIT Press.

Silberberg, S. (2013). *Places In The Making: How Placemaking Builds Places And Communities*. Cambridge, MA: MIT Press.

Sinnott, R. (1983). *Audiences, Acquisitions and Amateurs: Participation in the Arts in Ireland*. Dublin: Lansdowne Market Research/The Arts Council Ireland.

Soja, E. (1996). *Thirdspace: Journeys to Los Angeles and Other Real-And-Imagined Places*. Oxford: Basil Blackwell.

Stern, M. and Seifert, S. (2006). *Culture and Urban Revitalization: A Harvest Document*. University of Pennsylvania, PA: School of Social Work, Social Impact of the Arts Project.

UNESCO. (1972). Convention concerning the protection of the World Cultural and Natural Heritage. Available at: https://whc.unesco.org/en/about/. [Accessed: 25 March 2018].

Walker, J. and Marsh, S. (2018). 'A conversation between a collaborating artist and curator: Placemaking, socially engaged art, and deep investment in people', this volume.

Ward, S. V. (1998). *Selling Places: The Marketing and Promotion of Towns and Cities 1850–2000*. London: Routledge.

Weintraub, L. (2012). *To Life: Eco Art in Pursuit of a Sustainable Planet*. Berkley and Los Angeles, CA: University of California Press.

Wright, I. (2005). 'Place-Making as Applied Integral Ecology: Evolving an Ecologically-Wise Planning Ethic', *World Futures*, Vol. 61, pp. 127–137.

Index

Page numbers in *italic* indicate figures.

Printed in the United States
by Baker & Taylor Publisher Services